Signals and Communication Technology

The series “Signals and Communications Technology” is devoted to fundamentals and applications of modern methods of signal processing and cutting-edge communication technologies. The main topics are information and signal theory, acoustical signal processing, image processing and multimedia systems, mobile and wireless communications, and computer and communication networks. Volumes in the series address researchers in academia and industrial R&D departments. The series is application-oriented. The level of presentation of each individual volume, however, depends on the subject and can range from practical to scientific.

More information about this series at http://www.springernature.com/series/4748

Rohit M. Thanki • Vedvyas J. Dwivedi
Komal R. Borisagar

Multibiometric Watermarking with Compressive Sensing Theory

Techniques and Applications

Rohit M. Thanki
C. U. Shah University
Wadhwan City, Gujarat, India

Vedvyas J. Dwivedi
C. U. Shah University
Wadhwan City, Gujarat, India

Komal R. Borisagar
Atmiya Institute of Technology and Science
Rajkot, India

ISSN 1860-4862 ISSN 1860-4870 (electronic)
Signals and Communication Technology
ISBN 978-3-319-89239-9 ISBN 978-3-319-73183-4 (eBook)
https://doi.org/10.1007/978-3-319-73183-4

© Springer International Publishing AG 2018
Softcover re-print of the Hardcover 1st edition 2018
This work is subject to copyright. All rights are reserved by the Publisher, whether the whole or part of the material is concerned, specifically the rights of translation, reprinting, reuse of illustrations, recitation, broadcasting, reproduction on microfilms or in any other physical way, and transmission or information storage and retrieval, electronic adaptation, computer software, or by similar or dissimilar methodology now known or hereafter developed.
The use of general descriptive names, registered names, trademarks, service marks, etc. in this publication does not imply, even in the absence of a specific statement, that such names are exempt from the relevant protective laws and regulations and therefore free for general use.
The publisher, the authors and the editors are safe to assume that the advice and information in this book are believed to be true and accurate at the date of publication. Neither the publisher nor the authors or the editors give a warranty, express or implied, with respect to the material contained herein or for any errors or omissions that may have been made. The publisher remains neutral with regard to jurisdictional claims in published maps and institutional affiliations.

Printed on acid-free paper

This Springer imprint is published by Springer Nature
The registered company is Springer International Publishing AG
The registered company address is: Gewerbestrasse 11, 6330 Cham, Switzerland

Preface

With the invention of pattern recognition and automated human identification based on their biometric characteristics, a biometric-based system has become popular for the identification of an individual in many offices, institutions, and industries. This biometric-based system is implemented using various pattern recognition algorithms and signal processing algorithms. This system uses physiological and behavioral traits such as fingerprint, face, iris, signature, palm print, and speech for the identification of an individual. This system has good advantage like automatic identification of an individual over the knowledge-based system (i.e., password) and token-based system (i.e., identity card (ID)). But this system has some disadvantages such as noisy sensor data, intra-class variations, non-universality, and vulnerability against various attacks such as biometric data leakage or modification at the system storage and at the communication channel between two modules of the biometric-based system.

Nowadays, the multibiometric-based system is popular and widely used because it has overcome some disadvantages of the unique biometric-based system. This system has collected biometric data from multiple biometric sensors. In the multibiometric system, an identification of an individual is performed using multiple biometric characteristics. The two critical issues like designing of fusion technique and designing of security techniques are always associated when any multibiometric system is designed and implemented. The security issue for biometric data in the multibiometric system is addressed in this book. This book provides various techniques using digital watermarking and compressive sensing (CS) theory-based encryption for the security of biometric data in the multibiometric system. These presented techniques provide security to biometric data at the system storage and at the communication channel between two modules of the system.

This book introduces multibiometric watermarking technique with compressive sensing (CS) theory for the security of multiple biometric data in the multibiometric system. This book also introduces the application of compressive sensing theory in the biometric security field and biometric watermarking. In this technique, first, watermarked biometric data is converted into its encryption format using CS-based encryption process. This encrypted watermarked biometric data is inserted into other biometric data to generate watermarked biometric data. This watermarked biometric data represented secure watermark data by inserting one biometric data into other biometric data. For security checking and decision about the authentication of this watermarked biometric data, first, encrypted watermarked biometric data is detected from the watermarked biometric data. Then decryption of watermarked biometric data is performed using CS-based decryption process. This decrypted watermarked biometric data is compared with the original watermarked biometric data and makes the decision about the security and authenticity of the watermarked biometric data.

This book introduces four multibiometric techniques with CS theory: one technique for the protection of biometric data at the communication channel between two biometric data and three techniques for the protection of biometric data at the system storage of the multibiometric system. These watermarking techniques, namely, discrete wavelet transform (DWT), discrete cosine transform (DCT), singular-value decomposition (SVD), and fast discrete curvelet transform (FDCuT), are presented. These techniques which are classified based on transform coefficients of host biometric data are used for watermark embedding as DWT-based technique, DCT-based technique, SVD-based technique, and FDCuT-based technique.

The quality measures such as peak signal-to-noise ratio (PSNR) and structural similarity index measure (SSIM) are used for the performance analysis of all presented techniques. For the effect of these techniques on the performance of the multibiometric-based system, watermarked biometric data and recovered watermarked biometric data are used as multiple biometric data for the identification of an individual in this book. The quality measures such as the probability of verification, false rejection ratio (FRR), false acceptance rate (FAR), and equal error rate (EER) are used for the evaluation of the presented multibiometric system. The analysis of results of presented watermarking techniques shows that these techniques provide security to biometric data without affecting the performance of the multibiometric system. Also, the performance of the presented techniques is compared with that of existing watermarking techniques. The comparison of techniques shows that these presented techniques are outperformed by existing watermarking techniques in terms of security and perceptual transparency.

Contents

List of Figures

List of Tables

Chapter 1
Introduction

Abstract This chapter presents the background of a biometric system and its characteristics, its performance evaluation parameters, and its limitations. The basic overview of the multibiometric system and the motivation for the proposed research work is also given in this chapter.

1.1 Overview

Individual identity depends on various parameters (e.g., name, identity card) that are associated with an individual. The process of creating and maintaining individual identity is known as identity management (Nandakumar 2008). The method for the identity of the individual is called as an individual recognition or authentication. This process is a difficult task for any organization or institution. The three ways such as "something you know" (e.g., password), "something you carry" (e.g., ID card) and "something you are" (e.g., biometrics) are used for individual identification or authentication (Nandakumar 2008).

The individual identity based on knowledge-based system (e.g., password) and token-based system (e.g., ID card) can be easily manipulated or stolen by an impostor. The password can be easily guessed by an impostor using various passwords hacking software. The ID card can be duplicated by an impostor using various ID making software. Thus, these two cannot provide strong identification of an individual. Therefore, these two systems are not sufficient for the identification of an individual. Thus, researchers are introduced new identification system based on "something you are." This new identification system is called as a biometric system (Ratha et al. 2001; Jain et al. 2004; Nandakumar 2008; Jain et al. 2008; Jain and Kumar 2012; Ashbourn 2014; Harinda and Natgwirumugara 2015). Nowadays, an individual can be identified using this biometric system in institutions, offices, airports, and industries. The advantage of the biometric system is that it is more accurate compared to knowledge-based system and token-based system.

The term "biometrics" is coming from the two Geek word "bio" means "life" and "metrics" means "to measure" (Biometrics and Standards Report 2009).

© Springer International Publishing AG 2018

R. M. Thanki et al., *Multibiometric Watermarking with Compressive Sensing Theory*, Signals and Communication Technology,
https://doi.org/10.1007/978-3-319-73183-4_1

The biometric characteristics are divided into two types such as physiological and behavioral. These characteristics of an individual are used for identity of the individual in a biometric system (Biometrics and Standards Report 2009; Jain and Kumar 2012). The examples of physiological characteristics are fingerprint, face, ear, iris, teeth, and a sample of DNA. These characteristics are stable, unique, and lifetime. The examples of behavioral characteristics are speech, gait, keystroke, and signature (Jain et al. 2004; Biometrics and Standards Report 2009; Jain and Kumar 2012). These characteristics are dynamic and not a lifetime. The biometric trait is giving a link between an individual and his identity. The sample of biometric traits is called as a biometric template. These traits or templates cannot be easily stolen, forgotten, shared, or manipulated by an impostor (Jain et al. 2004; Nandakumar 2008; Jain and Kumar 2012). The templates of various biometric traits of an individual are shown in Fig. 1.1.

Any biometric traits of an individual have the following properties (Biometrics and Standards Report 2009).

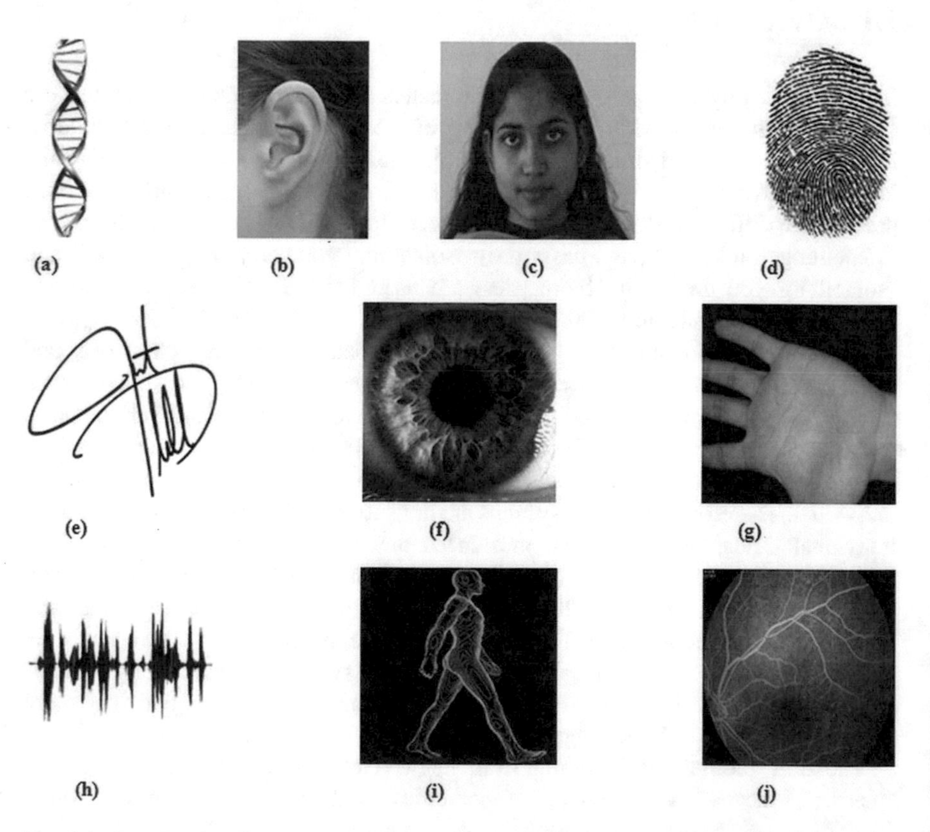

Fig. 1.1 Sample of different biometric traits of an individual. (**a**) DNA. (**b**) Ear. (**c**) Face. (**d**) Fingerprint. (**e**) Signature. (**f**) Iris. (**g**) Palm print. (**h**) Speech. (**i**) Gait. (**j**) Retina

Table 1.1 Comparison of various biometric traits (Jain et al. 2004)

Biometric Attributes	Fingerprint	Hand geometry	Iris	Signature	Face	Voice
Ease of use	Medium	High	Low	High	Medium	High
User acceptance	Medium	Medium	Low	High	High	High
Distinctiveness	High	Medium	High	Low	High	Low
Circumvention	Medium	Medium	Low	High	High	High
Long-term stability	High	Medium	High	Medium	Medium	Medium
Sensor cost	< $200	< $1500	< $400	< $300	< $100	< $25
Template size	0.5 KB	0.1 KB	256 bytes	0.2 KB	1 KB	2–3 KB

- *Universality:* Every individual should have its own biometric traits.
- *Distinctiveness:* Any two individuals should not have the same biometric traits.
- *Permanence:* The biometric traits should be invariant with time.
- *Collectability:* The biometric traits should be collected easily.

The designing criteria required for the any biometric system are performance, acceptability, exception handling, and cost for the system when it is used in real applications (Biometrics and Standards Report 2009). The comparison of various biometric traits is given in Table 1.1.

- *Performance:* This refers to the efficiency, accuracy, speed, and database required for implementation of the practical application.
- *Acceptability:* This refers to a biometric system that should be accepted by peoples who have used it.
- *Exception Handling:* This refers to a biometric system that should be easy to handle by peoples who do not know the operation of the system.
- *Cost:* This refers to cost that would be required when the biometric system is implemented in real-world application.

The comparison showed that template of face, fingerprint, and signature traits of an individual is widely used in the world. The reason behind using these traits is that these traits are unique to every individual, invariant, and easy to avail for acquisition (Jain et al. 2004; Biometrics and Standards Report 2009; Jain and Kumar 2012).

1.2 Biometric System

A simple biometric system is shown in Fig. 1.2a. This biometric system is known as a unimodal biometric system. This system used one biometric trait for the identity of an individual. This system is divided into five different modules such as a biometric sensor, a feature extractor, system database, matcher module, and decision module (Jain et al. 2004; Biometrics and Standards Reports 2009; Jain and Kumar 2012).

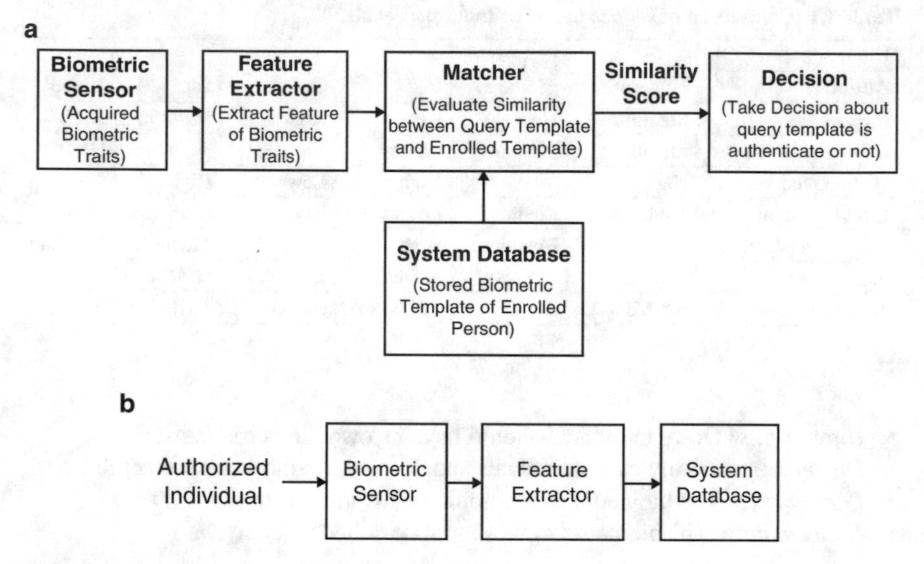

Fig. 1.2 (**a**) Biometric system. (**b**) Enrollment process

The first step of the system includes acquiring a biometric trait of an individual using an appropriate sensor. The salient features are extracted from the biometric trait using the feature extractor module. The operation of feature extractor module can be performed by the software-based algorithm. These extracted features in terms of the template are stored in system database with other identifying information such as a name or identification number (Jain and Kumar 2012). This process is known as an enrollment process and shown in Fig. 1.2b.

For the identity process, an individual presents the same biometric trait which is used during the enrollment process and features extracted using feature extractor module. These features in terms of the template are taken as a query template. This query template is compared with the stored template using the matcher module. This matcher module is given a similarity score between templates. Based on this similarity score, the system makes decision about the identity of an individual (Jain and Kumar 2012; Nandakumar 2008). This system can be operated in two modes such as verification and authentication (Biometric Testing and Statistics 2006; Jain et al. 2006; Jain et al. 2008; Jain and Kumar 2012).

1.2.1 Verification Operation

In this mode, the biometric system is verified by an authorized individual by comparing a submitted template with the previously enrolled template (Biometrics and Standards Report 2009; Biometrics Testing and Statistics 2006). During this operation, an individual makes a claim as to his/her identity. Then the biometric

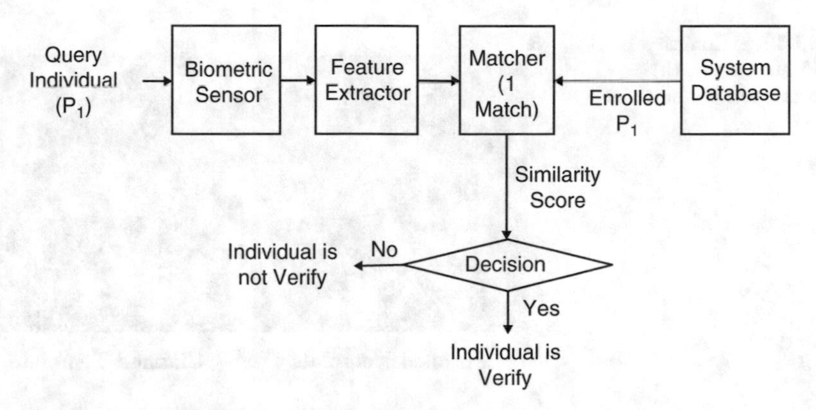

Fig. 1.3 Verification operation of biometric system

system determines if the individual's identity claim is true or false (Biometrics and Standards Report 2009; Biometrics Testing and Statistics 2006). Figure 1.3 shows a simple verification operation of the biometric system.

In this book, Euclidean distance-based biometric recognition method is used for verification of an individual. In this method, when the Euclidean distance between the claimed template and the enrolled template is near to zero, it indicated that the claimed template is the same as an enrolled template. This is called as a correct verification claim. If the Euclidean distance between the claimed template and enrolled template is high, it indicated that the claimed template is not the same as an enrolled template. This is called as an incorrect verification claim. Figure 1.4a shows a visual example of correct verification claim where a girl on the right makes a claim that she is an individual on the left. The Euclidean distance for this claim is zero. Figure 1.4b shows a visual example of incorrect verification claim. The Euclidean distance for this claim is 674.60.

The probability of verification of biometric system is 100% with FAR of 0% in the ideal case. But that is not possible, so system designer must analyze the system for various threshold values for a given application. Determining the threshold value is a difficult task because the probability of verification and FAR is not an independent value. If the verification rate increases, the FAR also increases (Biometrics Testing and Statistics 2006). The simple verification performance curve of a biometric system is shown in Fig. 1.5. It plots a curve for the probability of verification versus False Acceptance Rate (FAR) for different thresholds. The biometric system can be verified by different thresholds that move along the scale of 0–1 (Biometrics Testing and Statistics 2006).

In this book, verification of proposed watermarking technique-based multibiometric system is analyzed by setting various threshold values. The probability of verification performance of any biometric system is calculated using Eq. 1.1:

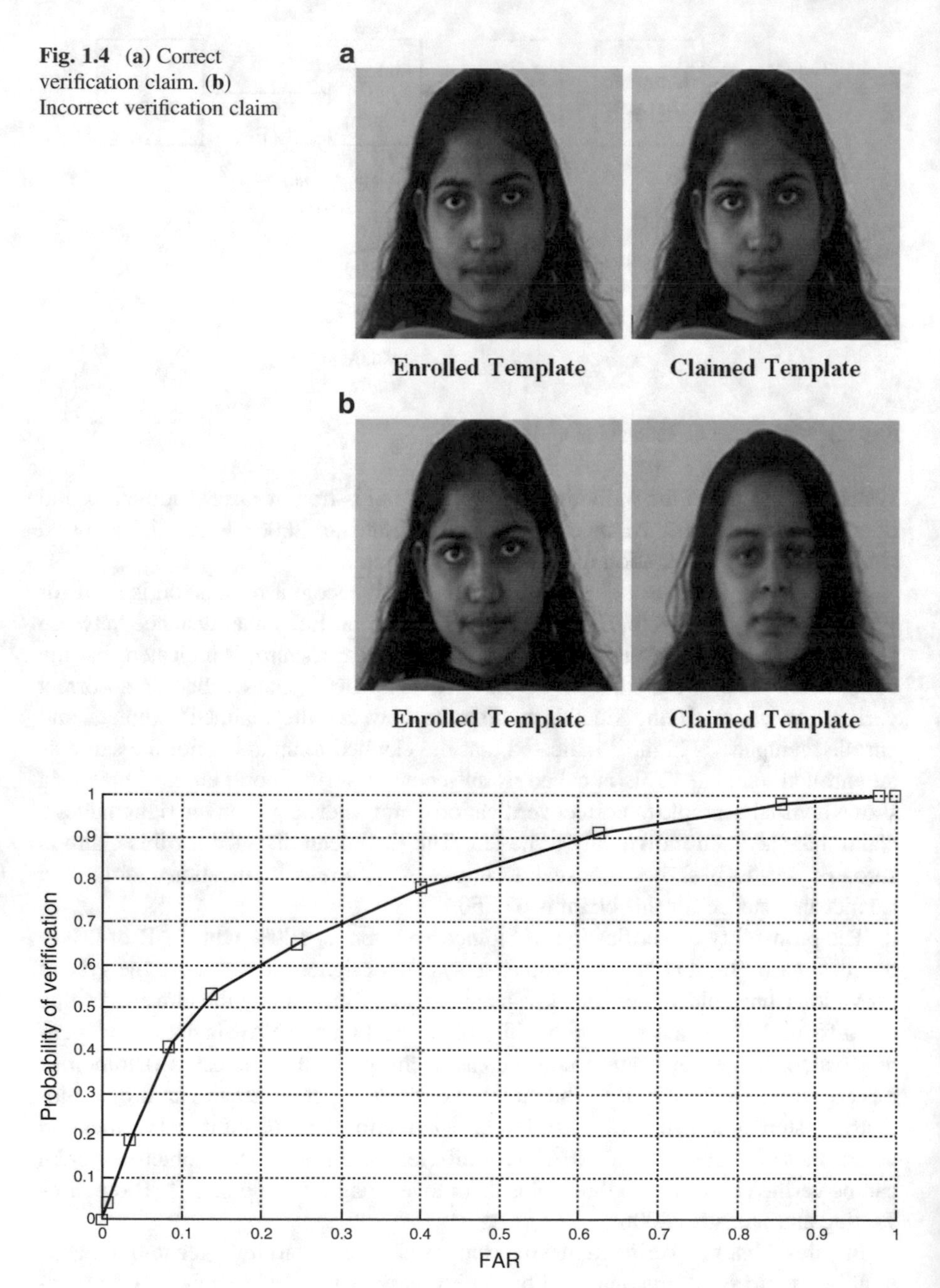

Fig. 1.4 (**a**) Correct verification claim. (**b**) Incorrect verification claim

Fig. 1.5 Verification performance curve of biometric system

$$V = \frac{(\text{No. of Matching Score}) < \text{Selected Threshold}}{\text{Total No. of Matching Score}} \tag{1.1}$$

where matching score = similarity score between authentic biometric image and enrolled biometric image in database and $V =$ probability of verification of system.

The probability of verification performance of the multibiometric system is calculated using Eq. 1.2. In this book, the multibiometric system is formed by using watermarked face-based system and reconstructed watermark fingerprint-based system:

$$V(\text{Multibiometric}) = \frac{V(\text{Face}) + V(\text{Fingerprint})}{2} \tag{1.2}$$

where V (face) = probability of verification of watermarked face-based system, V (fingerprint) = probability of verification of reconstructed watermark fingerprint-based system, and V (multibiometric) = probability of verification of multibiometric system.

1.2.2 Authentication Operation

In this mode, the biometric system attempts to find the identity of an individual. A biometric template is collected and compared with all the templates in a database (Biometrics and Standards Report 2009; Biometrics Testing and Statistics 2006). Figure 1.6a shows a simple authentication operation of the biometric system.

In this mode, the biometric system recognizes an authorized individual from entire enrolled individuals. It searches all templates stored in a system database for the identity of an individual. Figure 1.6b shows a visual example where an individual is compared with the entire enrolled individuals available at system database.

1.3 Performance Evaluation of Biometric System

The performance evaluation of any biometric system is done by using several statistical metrics. The performance of the biometric system can be evaluated by False Acceptance Rate (FAR), False Rejection Rate (FRR), Receiver Operating Characteristic (ROC) curve, and Equal Error Rate (EER). These rates are a function of threshold. The value of rates is obtained by operating the biometric system at various threshold values (Biometrics and Standards Report 2009; Jain and Kumar 2012; Giot et al. 2012).

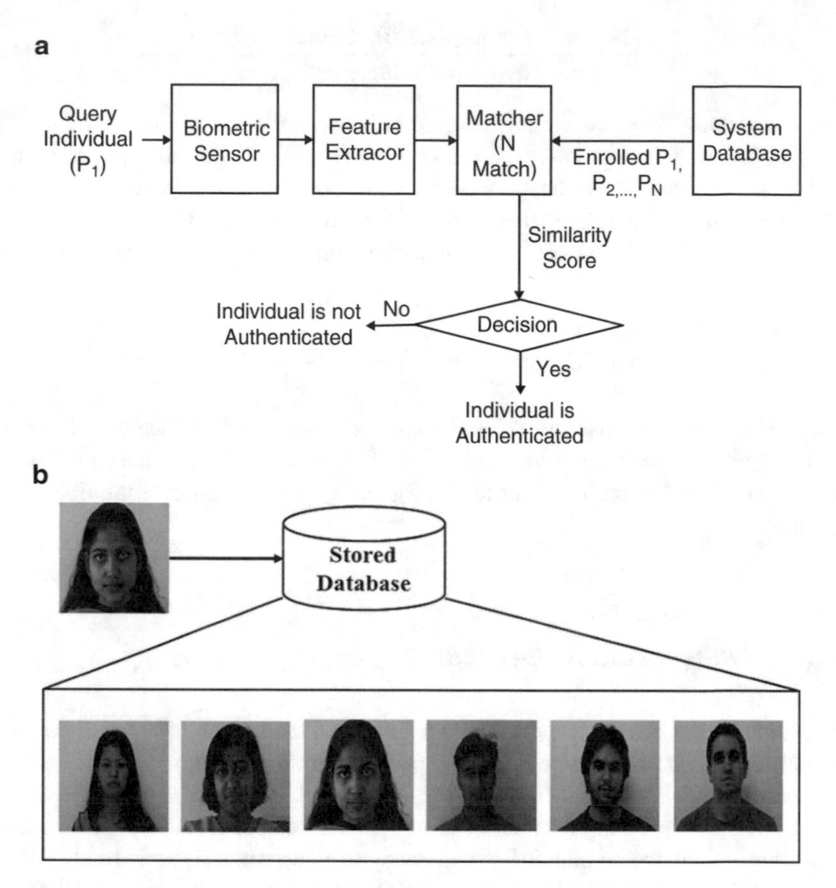

Fig. 1.6 (**a**) Authentication operation of biometric system. (**b**) Visual example of authentication operation

1.3.1 False Acceptance Rate (FAR)

A false acceptance occurs when an individual template is matched with another enrolled template. This is also known as False Match Rate (FMR) (Biometrics and Standards Report 2009). An example is that Rohit claims to be Jayesh, and then the system verifies this claim and allows Rohit to enter the system. The False Acceptance Rate (FAR) is calculated using Eq. 1.3 (Giot et al. 2012):

$$\text{FAR} = \frac{\text{No. of Matching Score} \leq \text{Selected Threshold}}{\text{Total No. of Matching Score}} \tag{1.3}$$

where FAR = False Acceptance Rate and matching score = similarity score between query biometric image and its closest match biometric image in database.

1.3.2 False Rejection Rate (FRR)

A false reject occurs when an individual template is not matched to its own enrolled template. This is also known as False Non-Match Rate (FNMR) (Biometrics and Standards Report 2009). An example is that Rohit claims to be Rohit, but the system denies the claim and does not allow Rohit to enter the system. The False Rejection Rate (FRR) is calculated using Eq. 1.4 (Giot et al. 2012):

$$\mathrm{FRR} = \frac{\text{No. of Matching Score} > \text{Selected Threshold}}{\text{Total No. of Matching Score}} \tag{1.4}$$

where FRR = False Rejection Rate and matching score = similarity score between query biometric image and its closest match biometric image in database.

1.3.3 Receiver Operating Characteristic (ROC) Curve

This curve is obtained by calculating the values of FAR, FRR for each tested threshold. It plots a curve for the FAR versus the FRR. This curve gives a trade-off between FAR and FRR values at various thresholds. This curve is also giving a trade-off between the system performance and security. Figure 1.7 shows a simple ROC curve for any biometric system (Biometrics and Standards Report 2009; Giot et al. 2012).

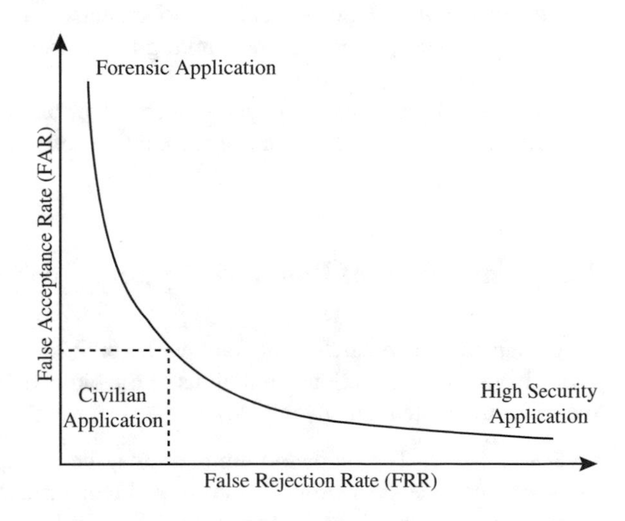

Fig. 1.7 ROC curve for biometric system

1.3.4 Equal Error Rate (EER)

It is the value on ROC curve where error rates of FAR and FRR have the same value (i.e., FRR = FAR). It is mostly used for comparison of biometric systems. The Equal Error Rate (EER) is calculated using Eq. 1.5:

$$\mathrm{EER} = \frac{\mathrm{FAR}_{\text{Selected Threshold}} + \mathrm{FRR}_{\text{Selected Threshold}}}{2} \tag{1.5}$$

where EER = Equal Error Rate, FAR = False Acceptance Rate, and FRR = False Rejection Rate.

FAR, FRR, and EER are playing an important role when designing any biometric system for any particular applications. When the biometric system is designed for the high security applications, then FAR must be as low as possible compared to FRR. When the biometric system is designed for the forensic applications, then FRR must be as low as possible compared to FAR. EER is used when the biometric system is designed for the civilian applications.

1.4 Applications of Biometric System

Biometric system is used in various applications. These applications are categorized into three main areas such as forensics, government, and commercial (Biometrics and Standards Report 2009; Jain and Kumar 2012; Giot et al. 2012).

- *Forensics Applications:* in criminal investigations
- *Government Applications:* personal documents, such as passports, ID cards and driving license, border and immigration control, social security, and e-governance
- *Commercial Applications:* physical access control, network logins, ATM, credit card, mobile phones, PDA, facial recognition software, and e-health

1.5 Limitations of Biometric System

A. K. Jain and his research team (Jain et al. 2006; Jain et al. 2008; Jain and Kumar 2012) have given the various limitations of the biometric system. The limitations of the biometric system are given below:

- *Sensor Noise:* The biometric template may be corrupted and destroyed by the noise present at the sensor. This corrupted biometric template may be incorrectly matched with enrolled templates (Fig. 1.8), which then produce a result in such a way that an authorized person is being incorrectly rejected by the system.

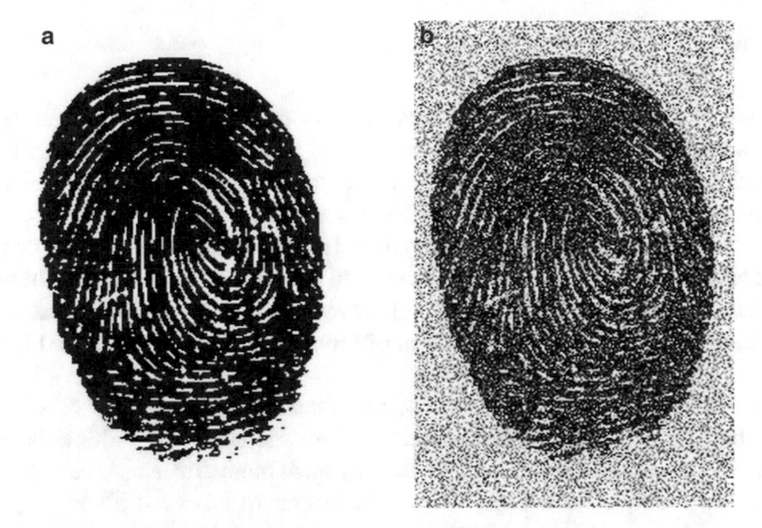

Fig. 1.8 (a) Original biometric trait. (b) Noisy biometric trait

Fig. 1.9 (a–c) Intra-class variations

- *Intra-class Variations:* The biometric template of an individual is acquired at various conditions during the authentication process. These templates are different than template generated during the enrollment process. This variation is created due to incorrect interaction and modification of biometric sensor (Fig. 1.9).
- *Distinctiveness:* Biometric traits are expected to differ for individuals, but there are large interclass similarities in the features of these traits.
- *Nonuniversality:* When an individual's biometric trait is acquired, then the biometric system cannot perform any operation on the particular biometric trait.
- *Spoof Attacks:* An unauthorized individual may be trying to create spoof biometric characteristics of enrolled data of an authorized individual.

1.6 Multibiometric System

The unimodal biometric system is vulnerable to different attacks such as fake or spoof biometric template at system database and template stone over the communication channel (Ratha et al. 2001). To overcome these limitations, a new biometric system is introduced by A. K. Jain and his research team (Jain et al. 2004; Jain et al. 2004; Nandakumar 2008). This new system is known as multibiometric or multimodal biometric system. This biometric system used two or more than two biometric traits for the identity of an individual. This system reduces problem-related large population coverage and spoof detection of biometric data using multiple biometric traits.

There are many types of multibiometric system such as multiple sensors (e.g., taking biometric templates using different biometric sensors), multiple biometric templates (e.g., takes two or more than two different biometric templates of the same individual), multiple units (e.g., two or more fingerprints of different fingers of individual), and multiple matching (e.g., matches individual biometric templates using different matching algorithms) (Theime 2003; Ross and Jain 2004; KO 2005; Jain et al. 2006; Agrawal 2007; Lupu and Pop 2008; Kaur et al. 2010; Sasidhar et al. 2010; Panchal and Singh 2013; Gayathri and Rani 2013). The various multibiometric systems are shown in Fig. 1.10.

The multibiometric system is operated in three different modes such as serial, parallel, and hierarchical (Theime 2003; Ross and Jain 2004; KO 2005; Jain et al. 2006; Agrawal 2007; Lupu and Pop 2008; Kaur et al. 2010; Sasidhar et al. 2010;

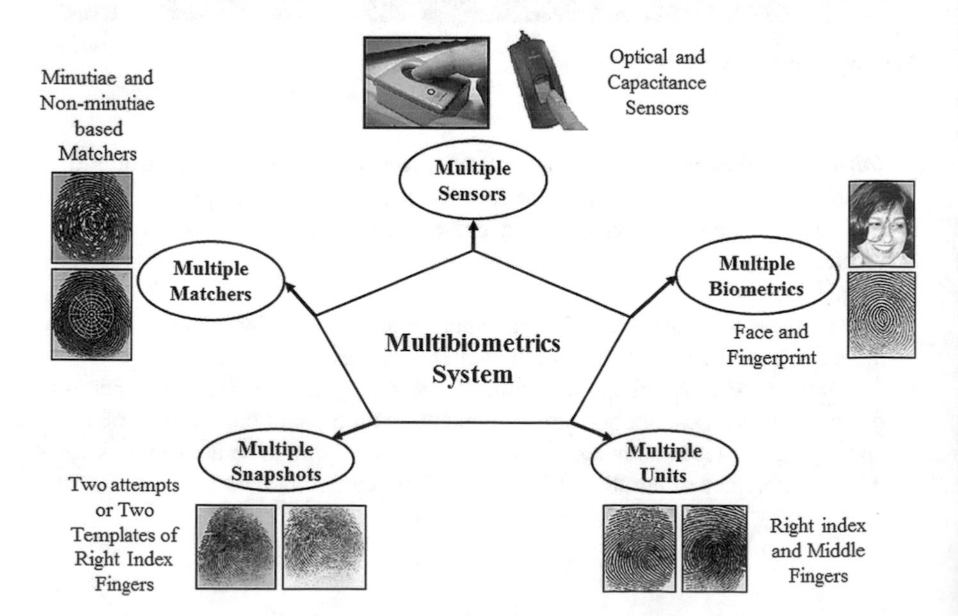

Fig. 1.10 Various multibiometric systems (Jain et al. 2004; Ross and Jain 2004)

Panchal and Singh 2013; Gayathri and Rani 2013). In serial mode of operation, the identity of the individual is performed using the first biometric trait. If the identity of an individual cannot be possibly performed, then the second biometric trait is extracted from the first biometric trait. The identity of an individual can be performed by using the second biometric trait. In parallel mode of operation, the identity of the individual can be performed using both biometric traits simultaneously. In the hierarchical mode of operation, the identity of an individual can be performed using a tree structure.

In the multibiometric system, multibiometric traits are generated using various fusion approaches (Ross and Jain 2004; Jain et al. 2004; Jain et al. 2006). The different fusion approaches for a multibiometric system are shown in Fig. 1.11 (Ross and Jain 2004; Jain et al. 2004; Jain et al. 2006).

- *Sensor Level:* Using multiple sensors, multiple traits of an individual are acquired, and these traits are fused to get multiple traits of an individual.
- *Feature Level:* In this fusion approach, features of two or more than two traits of an individual are fused to generate new features for multiple traits of an individual. This level of fusion is shown in Fig. 1.11a.
- *Match Score Level:* At this level of fusion, multiple matching scores are produced by matchers which are used to generate a single score. Based on this single score, the identity of the individual is performed. This level of fusion is shown in Fig. 1.11b.
- *Decision Level:* In this level of fusion, fused decision of individual matching module created one decision value. Based on newly decision value, the identity of the person is performed. This level of fusion is shown in Fig. 1.11c.

The multibiometric system has more advantages compared to the unimodal biometric system. But, there are two research problems associated with the multibiometric systems such as the designing of security techniques and designing of fusion techniques. In this book, various security techniques are presented for biometric data protection in the multibiometric system. These security techniques are implemented using compressive sensing (CS) theory and watermarking.

1.7 Motivation for Book

The biometrics plays a very important role in the identity of the person in many applications such as security of personal documents, ID cards, driving license, and access to mobile phones and laptops. The privacy and security concerns of biometric data against modification or spoof attack are always present in any biometric system. The examples of spoofed fingerprint and spoofed face are given in Fig. 1.12. The major drawback of any biometric system is that protection is required for biometric image or template against attack at various points of the biometric system. The attacks on biometric system are described in Chap. 3.

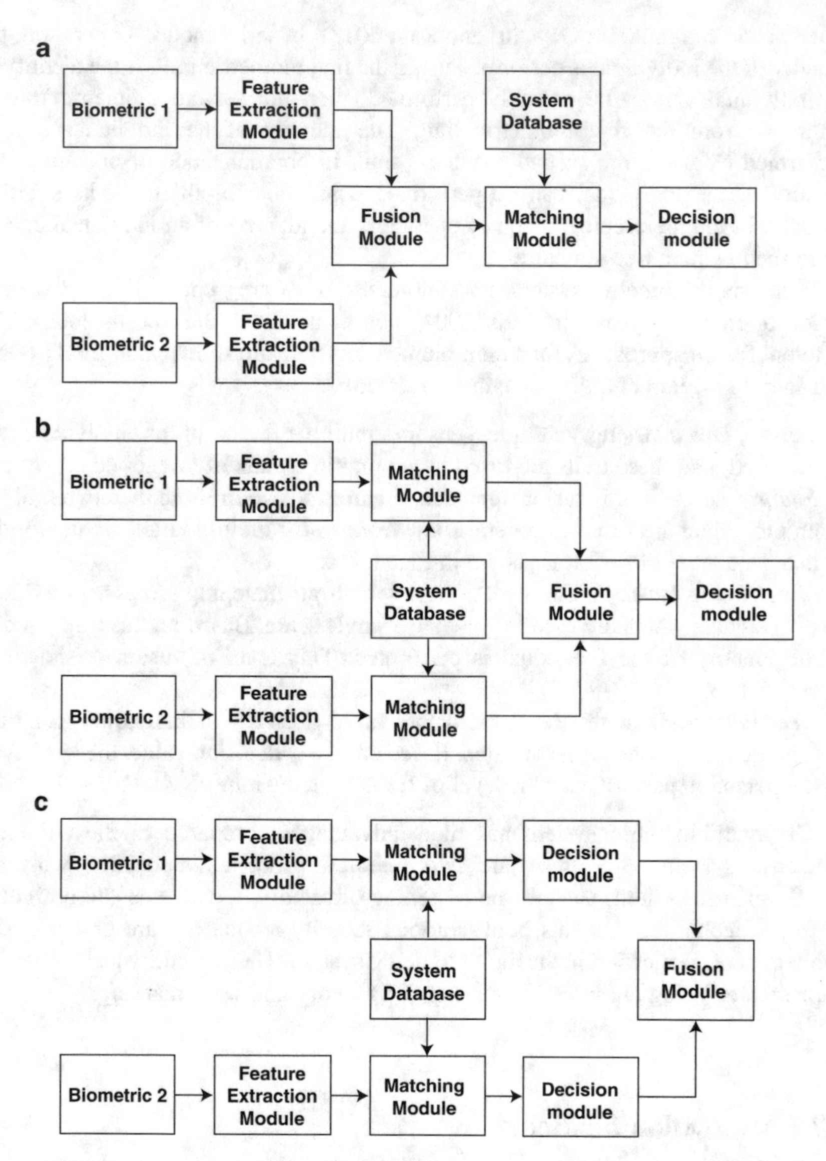

Fig. 1.11 Various fusion approaches in multibiometric systems (Jain et al. 2004; Ross and Jain 2004). (**a**) Fusion at feature extraction level. (**b**) Fusion at matching score level. (**c**) Fusion at decision level

Compressive sensing (CS) theory and watermarking are not new methods for biometric contents. While watermarking is used for security of biometric data, compressive sensing (CS) is used for compression as well as encryption of biometric data.

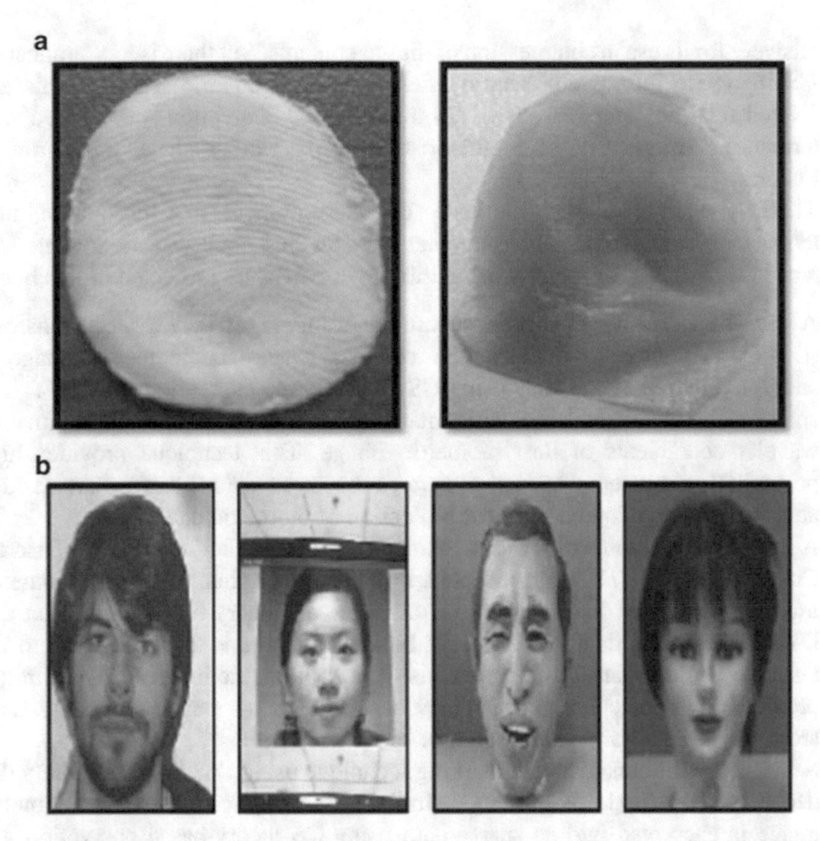

Fig. 1.12 Spoofed biometric modalities (Akhtar 2012; Zhang et al. 2011). (**a**) Spoofed fingerprint examples: silicone and gelatin. (**b**) Spoofed face examples: photo, video replay, rubber, and silica gel

A lot of watermarking techniques have been designed and implemented for protection of biometric data such as fingerprint, iris, facial, and palm print. In literature, the very less watermarking techniques are designed for the authentication of multibiometric traits. The existing biometric watermarking techniques are designed using various transforms like Discrete Cosine Transform (DCT), Discrete Wavelet Transform (DWT), and so on. The problem related to existing watermarking techniques is that watermark biometric data is directly embedded into host biometric data. Thus, an attacker can easily extract watermark biometric data from watermarked biometric data if an attacker gets correct extraction algorithm and secret key. Another problem related to existing watermarking techniques is that the effect of these techniques on the performance of the multibiometric system is missing. Also, these existing watermarking techniques have very less perceptual transparency and payload capacity.

The compressive sensing is new signal process theory. This theory acquires a signal in a compressed as well as encrypted manner. CS theory has been studied extensively for its use in compression of digital contents. But there is less application of CS theory to information security field such as biometric security. This gap motivated for study and analysis of CS theory based encryption is combined with watermarking for security of biometric data using various signal transforms in this book.

These above limitations motivate the development and design of new watermarking techniques, which are presented in this book. The following four advanced techniques for security of multibiometric data are presented in this book.

- A multibiometric watermarking technique using Discrete Wavelet Transform (DWT) is presented. Initially, the original watermark biometric image is encrypted into its sparse data using CS theory-based encryption and DWT. The sparse data of watermark biometric image is then inserted into the approximation wavelet coefficients of host biometric image. This technique provides high perceptual transparency as well as better robustness against basic watermarking attacks. This technique is used for protection of biometric data.
- A hybrid multibiometric watermarking technique using DWT and Discrete Cosine Transform (DCT) is presented. Initially, the original watermark biometric image is encrypted into its sparse data using CS theory-based encryption and DWT. The sparse data of watermark biometric image is then inserted into the Discrete Cosine Transform coefficients of host biometric image. This technique provides high perceptual transparency as well as better payload capacity. This technique is used for authentication of biometric data.
- A hybrid multibiometric watermarking technique using DWT and singular value decomposition (SVD) is presented. Initially, the original watermark biometric image is encrypted into its sparse data using CS theory-based encryption and DWT. The SVD is applied on sparse data of watermark biometric image to get its singular value. These singular values of watermark biometric image are inserted into singular value of details wavelet coefficients of host biometric image. This technique is used for authentication of biometric data.
- A hybrid multibiometric watermarking technique using Fast Discrete Curvelet Transform (FDCuT) and DCT is presented. Initially, the original watermark biometric image is encrypted into its sparse data using CS theory-based encryption and DCT. The sparse data of watermark biometric image are inserted into high-frequency curvelet coefficients of host biometric image. This technique provides high perceptual transparency as well as better payload capacity. This technique is used for authentication of biometric data.

1.8 Book Organization

This chapter briefly discussed general characteristics of biometric system and multibiometric system. In addition, the motivation behind writing this book is presented. The rest of this book is organized as follows.

Chapter 2 presents related works available in the literature. Chapter 3 gives a proposed research methodology and resources used in the present work. Chapter 4 presents information and mathematics of various terminologies used in implementation of the present work.

Chapter 5 gives a DWT-based multibiometric watermarking technique and its performance analysis. Chapter 6 gives a DWT-DCT-based multibiometric watermarking technique and its performance analysis. Chapter 7 gives a DWT-SVD-based multibiometric watermarking technique and its performance analysis. Chapter 8 gives a DCT-FDCuT-based multibiometric watermarking technique and its performance analysis. Chapter 9 concludes this book with some future research directions.

References

Agrawal, M. (2007). *Design approaches for multimodal biometric system*. M. Tech. thesis, Department of Computer Science and Engineering, IIT, Kanpur.

Akhtar, Z. (2012). *Security of multimodal biometric systems against spoof attacks*. Ph.D. thesis, University of Cagliari.

Ashbourn, J. (2014). *Biometrics in the new world: The cloud, mobile technology and pervasive identity*. Cham: Springer.

Biometrics and Standards. (2009). *ITU-T Technology Watch Report, December*. Available http://www.itu.int/dms_pub/itu-t/oth/23/01/T230100000D0002MSWE.doc

Biometrics Testing and Statistics. (2006). *National Science and Technology Council (NSTC) Report*. Available www.biometrics.gov/documents/biotestingandstats.pdf

Gayathri, D., & Rani, R. (2013). Multimodal biometric system: An overview. *International Journal of Advanced Research in Computer and Communication Engineering, 2*(1), 898–902.

Giot, R., El-Abed, M., & Rosenberger, C. (2012). Fast computation of the performance evaluation of biometric systems: Application to multibiometrics. *Future Generation Computer Systems, 1*, 1–30.

Harinda, E., & Natgwirumugara, E. (2015). Security & privacy implications in the placement of biometric-based ID card for Rwanda Universities. *Journal of Information Security, 6*, 93–100.

Jain, A., & Kumar, A. (2012). Biometric recognition: An overview. In E. Mordini & D. Tzovaras (Eds.), *Second Generation Biometrics: The Ethical, Legal and Social Context* (pp. 49–79). Dordrecht: Springer.

Jain, A., Ross, A., & Prabhakar, S. (2004). An introduction to biometric recognition. *IEEE Transactions on Circuits and Systems for Video Technology, Special Issue on Image and Video Based Biometrics, 14*(1), 4–20.

Jain, A., Ross, A., & Pankanti, S. (2006). Biometrics: A tool for information security. *IEEE Transactions on Information Forensics and Security, 1*(2), 125–143.

Jain, A., Nandakumar, K., & Nagar, A. (2008). Biometric template security. *EURASIP Journal on Advances in Signal Processing, Special Issue on Advanced Signal Processing and Pattern Recognition Methods for Biometrics*, January, pp. 1–17.

Kaur, M., Girdhar, A., & Kaur, M. (2010). Multimodal biometric system using speech and signature modalities. *International Journal of Computer Applications, 5*(12), 13–16.

Ko, T. (2005). Multimodal biometric identification for large user population using fingerprint, face and iris recognition. *Proceeding of the IEEE 34th Applied Imagery and Pattern Recognition Workshop (AIPR05)*.

Lupu, E., & Pop, P. (2008). Multimodal biometric systems overview. *ACTA Technica Napocensis, 49*(3), 39–44.

Nandakumar, K. (2008). *Multibiometric systems: Fusion strategies and template security*. Ph.D. thesis, Michigan State University, USA.

Panchal, T., & Singh, A. (2013). Multimodal biometric system. *International Journal of Advanced Research in Computer Science and Software Engineering, 3*(5), 1360–1363.

Ratha, N., Connell, J., & Bolle, R. (2001). Enhancing security and privacy in biometric based authentication systems. *IBM Systems Journal, 40*(3), 614–634.

Ross, A., & Jain A. (2004). Multimodal biometrics: An overview. *Proceedings of 12th European Signal Processing Conference (EUSIPCO)*, pp. 1221–1224.

Sasidhar, K., Kakulapati, V., Ramakrishna, K., & Ka, K. (2010). Multimodal biometric systems – study to improve accuracy and performance. *International Journal of Computer Science & Engineering Survey (IJCSES), 1*(2), 54–61.

Theime, M. (2003). *Multimodal biometric systems: Applications and usage scenarios*. Arlington: Biometric Consortium Conference.

Zhang, Z. Yi, D., Lei, Z., & Li, S. (2011). *Face liveness detection by learning multispectral reflectance distributions*. IEEE International Conference on Automatic Face and Gesture Recognition, pp. 436–441.

Chapter 2
Background Information and Related Works

Abstract This chapter presents literature review in context to existing works with an aim of identifying the present state of research in this domain. This chapter also presents background information regarding biometric watermarking technique and compressive sensing (CS) theory-based encryption technique. The multibiometric watermarking technique is one of the types of biometric watermarking technique. This chapter described application of CS theory in information security field.

2.1 Background Information

This section presents technical details on biometric watermarking technique and CS theory-based encryption technique. This section gives various properties of watermarking and CS theory. This section also explained the quality measures used for performance check of watermarking technique.

2.1.1 Biometric Watermarking Technique

In the last few decades, information such as images, videos, audio, books, and documents are transferred on the Internet. This information is easy to download without any permission. This situation creates the problem of copyright protection and authentication. So researchers have described a new technique which is known as a watermarking. The watermarking technique provides security for copyright protection, copyright authentication, and temper identification. Watermarking is the existence of steganography, which is a technique to hide the data from an unauthorized individual. The watermarking has overcome the limitation of steganography technique such as point-to-point communication and lack of security against any changes. The watermarking technique is used for protection of any data over any communication media. Digital watermarking is a technique to embed a watermark data into a host medium in such a manner that the common

© Springer International Publishing AG 2018
R. M. Thanki et al., *Multibiometric Watermarking with Compressive Sensing Theory*, Signals and Communication Technology,
https://doi.org/10.1007/978-3-319-73183-4_2

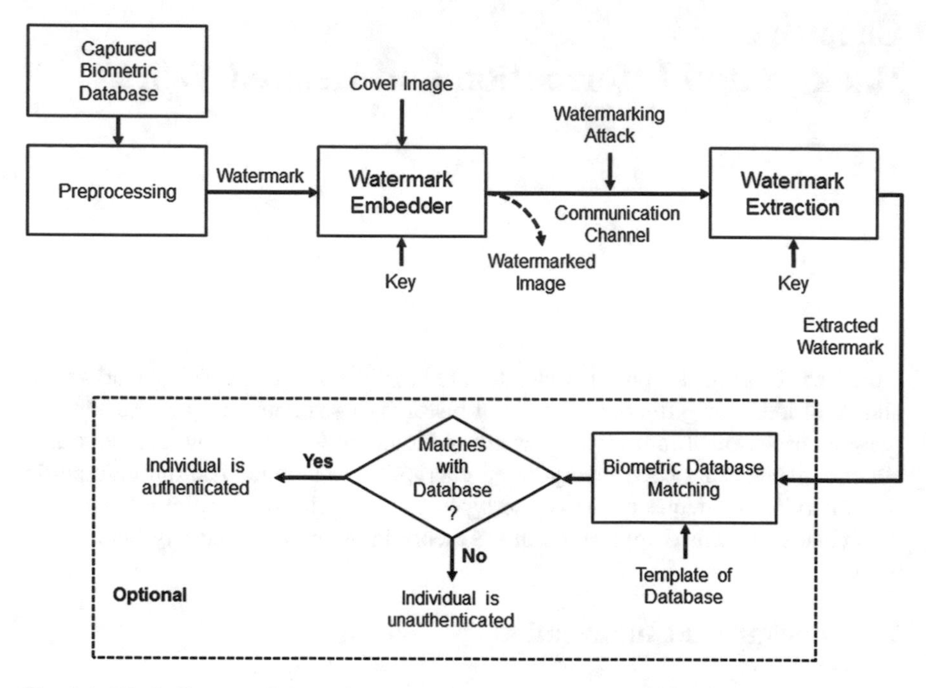

Fig. 2.1 Block diagram of biometric watermarking technique (Rege 2012)

man cannot visualize the watermark with a naked eye (Langelaar et al. 2000; Cox et al. 1997).

Biometric data is used as watermark information which is embedded into a digital content, and it is called as a biometric watermarking. When biometric data is used as watermark information which is embedded into another biometric data, it is called as a multibiometric or multimodal biometric watermarking. A biometric watermarking system is viewed as a communication system which has three main elements such as an embedder, a communication channel (which is optional), and extraction (Rege 2012). The block diagram of the general biometric watermarking system is shown in Fig. 2.1.

The watermark embedder embedded the biometric data as a watermark into the host image. The input of watermark embedder is a host image, watermark biometric image, and secret key. The output of watermark embedder is watermarked image. This watermarked image is transmitted through wire or wireless communication. On the watermark extraction side, the extraction has extracted watermark biometric image from watermarked image using a secret key. The input of watermark extraction is watermarked image and secret key. The output of watermark extraction is extracted watermark biometric image.

The extracted watermark biometric image is compared with original watermark biometric image for a decision about the authenticity of an individual. If the result of the matching of the watermark biometric image is greater than the selected threshold, then an individual is authenticated; otherwise an individual is unauthenticated.

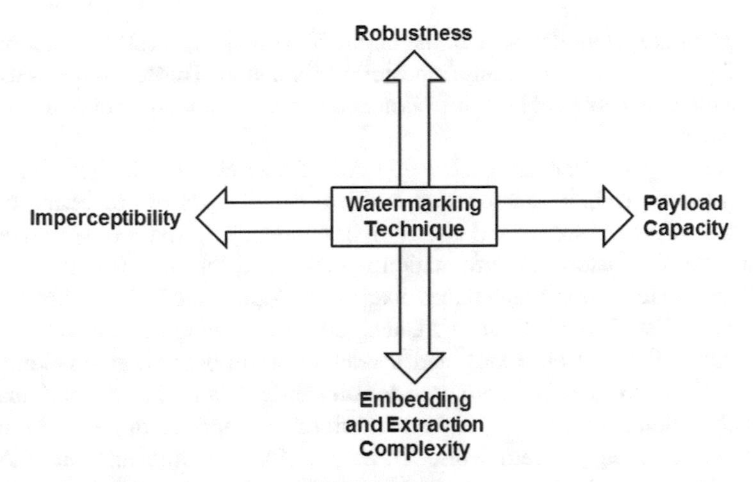

Fig. 2.2 Requirements of biometric watermarking technique

2.1.1.1 Requirements of Biometric Watermarking Technique

The requirements of biometric watermarking technique are same as requirements of digital watermarking. These requirements are described below (Kothari 2013; Thanki et al. 2011; Langelaar et al. 2000). All biometric watermarking technique is distinguished based on these parameters which are shown in Fig. 2.2.

- *Robustness*: The watermarking techniques can survive for different modification to the watermarked content. The watermarked content should not be destroyed by ADC conversion, DAC conversion, cropping, resampling, rotation, quantization, scaling, noise addition, and compression attacks.
- *Imperceptibility*: When the biometric template is embedded in a host medium such that biometric template could not visible by the human visual system (HVS).
- *Payload Capacity*: It is the size of the biometric template that is embedded into a host medium. It depends on the technique used for the watermarking.
- *Embedding and Extraction Complexity*: This distinguishes between different watermarking techniques based on complexity and security.

2.1.1.2 Types of Biometric Watermarking Technique

The various types of biometric watermarking techniques are available in the literature. The various categories of biometric watermarking technique are given below (Rege 2012):

- *Depending on Authenticity of Biometric Data*: The biometric watermarking technique is divided into two types based on the authentication whether fragile or robust. The fragile biometric watermarking technique could not survive by any impostor manipulation. This technique is used for biometric image or template protection at system database of the biometric system. The robust watermarking technique could survive by any impostor manipulation. This technique is used for biometric image or template protection at communication channel of the biometric system.

- *Dependent on Processing Domain of Biometric Data*: The biometric watermarking technique is divided into two major types of processing domain whether a space or spatial and transform or frequency. In space or spatial domain technique, the watermark information is embedded into the host image. Three different watermarking techniques like Least Significant Bit substitution technique (Bedi et al. 2011; Chan and Cheng 2004), correlation-based watermarking technique (Vatsa et al. 2004), and spread spectrum-based watermarking technique (Kim and Lee 2009) are used for biometric data in the spatial domain. In transform domain technique, various transforms are applied to the host image to find out frequency coefficients. These transform coefficients are modified according to watermark biometric information. Most common transform used in watermarking techniques are Discrete Cosine Transform (DCT) (Paunwala and Patnaik 2014; Behera and Govindan 2013; Isa and Aljareh 2012; Bedi et al. 2012; Naik and Holambe 2010; Mathivadhani and Meena 2010; Vatsa et al. 2004), Discrete Wavelet Transform (Chaudhary et al. 2013; Mathivadhani and Meena 2010; Noore et al. 2007a, b; Vatsa et al. 2009, Vatsa et al. 2005; Vatsa et al. 2005), and Singular Value Decomposition (Chaudhary et al. 2013; Joshi et al. 2013; Inamdar and Rege 2012).

2.1.1.3 Applications of Biometric Watermarking Technique

The biometric watermarking is applicable in below applications:

- *Ownership Identification*: An authenticated person can extract the watermark from his/her content to prove his/her ownership.
- *Fingerprinting*: A person can embed a fingerprint as a watermark into his/her content in such a way that it is difficult to remove. This allows the authenticated person to trace impostor if the data is distributed illegally.
- *Authentication*: The authenticated person can embed a fragile biometric watermark into the host biometric data to provide a security to biometric data. If any tampering is applied on the watermarked content then it destroys the fragile watermark content.
- *Visible Watermarking*: A visible biometric watermark such as a signature can be put on documents to prevent the creation of illegal copies of the document.

2.1.1.4 Performance Evaluation of Biometric Watermarking Technique

The performance of watermarking technique is evaluated using various quality measures (Wang and Bovik 2004; Voloshynovskiy et al. 2001; Petitcolas 2000;

Wolfgang et al. 1999; Kutter and Petitcolas 1999). These quality measures are given by Eqs. 2.1, 2.2, and 2.3. These quality measures have given how much change has taken place in the host image due to the embedding of the biometric watermark image.

$$\mathrm{MSE} = \frac{1}{M \times N} \sum_{x=1}^{M} \sum_{y=1}^{N} (I(x,y) - \mathrm{IW}(x,y))^2 \quad (2.1)$$

$$\mathrm{PSNR} = 10 \times \log \frac{255^2}{\mathrm{MSE}} \quad (2.2)$$

where MSE is the mean square error, PSNR is the peak signal to noise ratio, $I\,(x, y)$ is the original host image, and IW (x, y) is the watermarked image.

$$\mathrm{SSIM}(x,y) = \frac{(2xy + c_1)(2\sigma_{xy} + c_2)}{(\hat{x}2 + \hat{y}2 + 1)\left(\sigma_x^2 + \sigma_y^2 + c_2\right)} \quad (2.3)$$

where x and y are corresponding windows of the same size of the original watermark biometrics and reconstructed watermark biometric images and $\hat{x}$ and $\hat{y}$ are the corresponding averages of x and y, respectively, and σ_x^2 and σ_y^2 are the corresponding variances of x and y. σ_{xy} is the covariance of x and y and c_1 and c_2 are appropriate constants.

The abbreviation PSNR is called as a peak signal to noise ratio. It is used to measure the difference between a host biometric data and its watermarked version at embedder side (Voloshynovskiy et al. 2001; Petitcolas 2000; Wolfgang et al. 1999; Kutter and Petitcolas 1999). The equation of PSNR includes the measurement of MSE. Mean square error (MSE) finds the difference between a host biometric data and its watermarked version. The MSE is a measured value in the general scale, while PSNR is a measured value in the logarithmic scale.

At the extraction side, the abbreviation SSIM is called as a structural similarity index measure. It is the generalized form of the Universal Image Quality Index (UQI). It is used to find similarity between an original watermark biometric data and its reconstructed version (Bedi et al. 2012; Wang and Bovik 2004).

2.1.1.5 Watermarking Attacks

The aim of designing any biometric watermarking technique is to provide security to biometric data against impostor manipulation. When watermarked data is transmitted to different modules of the biometric system, then it may be corrupted by an attacker. The attacker is trying to remove or corrupt biometric data in the biometric system. Some of the watermarking attacks (Voloshynovskiy et al. 2001; Petitcolas 2000; Wolfgang et al. 1999; Kutter and Petitcolas 1999) are given below. These attacks are used for analysis of performance of proposed multibiometric techniques in this book.

- *Filtering Attacks*: There are different types of image filters such as median filter, mean filter, and Gaussian low-pass filter that are applied to watermarked biometric image. In this thesis, mask sizes of all three filters are chosen as 3×3, 5×5, and 7×7.
- *Additive Noise Attacks*: There are different types of noise such as Gaussian noise, salt and pepper noise, and speckle noise that are applied on watermarked biometric data. The Gaussian noise PDF is given as (Gonzales and Woods 2002)

$$p(z) = \frac{1}{\sqrt{2\pi\sigma}} e^{-(z-\mu)^2/2\sigma^2} \tag{2.4}$$

where z represents the intensity, μ represents the mean (average) value of z, σ is the standard deviation, and σ^2 is the variance of z.

The salt and pepper noise PDF is given as (Gonzales and Woods 2002)

$$\begin{aligned} p(z) &= Pa, z = a \\ p(z) &= Pb, z = b \\ P(z) &= 0, \text{otherwise} \end{aligned} \tag{2.5}$$

The speckle noise inherently exists and degrades the quality of watermarked biometric image.

In this book, Gaussian noise is generated using the mean value of 0 and variance value of 0.0001. Salt and pepper noise is generated using the variance value of 0.0005. Speckle noise is generated using the variance value of 0.0004. The performance of proposed multibiometric watermarking technique is evaluated after adding these three noises into the watermarked biometric image.

- *JPEG Compression Attack*: The watermarked biometric image is compressed by the JPEG compression. In this thesis, quality factor Q is taken from a range of 90–50.
- *Histogram Equalization Attack*: The histogram equalization operation modified and equalized the gray level of watermarked biometric image. The performance of proposed watermarking technique is measured after the application of this attack.
- *Geometric Attacks*: The basic geometric attacks like flipping, rotation, and cropping are applied on watermarked biometric image. The left-right flipping, rotation with 90°, and 20% cropping are applied on watermarked biometric image in this thesis. The performance of proposed watermarking technique is measured after the application of this attack.
- *Other Advanced Attacks*: The advanced watermarking attacks like sharpening and blurring are also applied on watermarked biometric image. The performance of proposed watermarking technique is measured after the application of these attacks.

2.1.2 CS Theory-Based Encryption Technique

To overcome the limitation of Shannon-Nyquist sampling criteria in existing compression approaches, Candes and Donoho introduced new mathematical theory for signal acquisition and compression (Donoho 2006; Candes 2006). This mathematical theory is known as "compressive sensing or sampling (CS)" theory. This theory is equally applicable for digital images. This theory states that the image can successfully recover from its few sparse data (Donoho 2006; Candes 2006; Baraniuk 2007). A necessary condition for CS theory is that image must be sparse in its own domain. The beauty of this theory is that it simultaneously compresses and encrypts image based on measurement matrix, which may have binary, Fourier, or Gaussian nature. The details of image encryption and decryption based on CS theory are shown in Fig. 2.3.

2.1.2.1 CS Theory-Based Encryption Process

In CS encryption process, the image is converted into its sparse measurements using sparse coefficients and measurement matrix. The image in terms of its sparse measurements is called as compressed and encrypted image. The CS theory-based encryption process for grayscale standard image is shown in Fig. 2.4.

The steps for CS encryption process are given as follows:

- Convert image into its sparse coefficients using image transform basis matrix as follows:

$$x_{N \times N} = \Psi_{N \times N} \cdot f_{N \times N} \cdot \Psi'_{N \times N} \tag{2.6}$$

where x is sparse coefficients of the image, f is the original image, and Ψ is transform basis matrix.

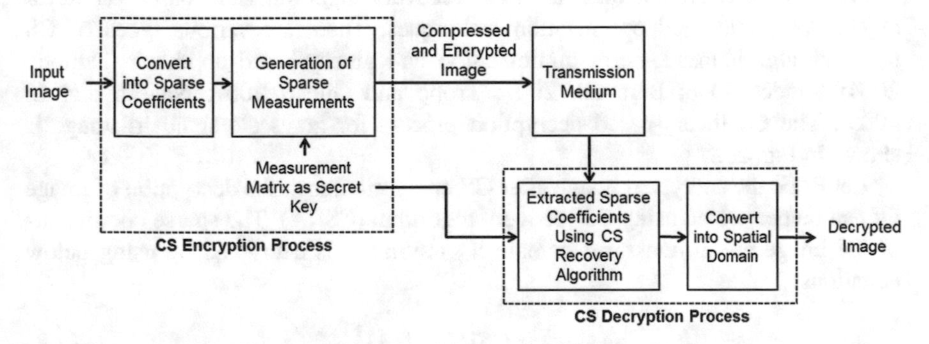

Fig. 2.3 CS theory-based image encryption and decryption process

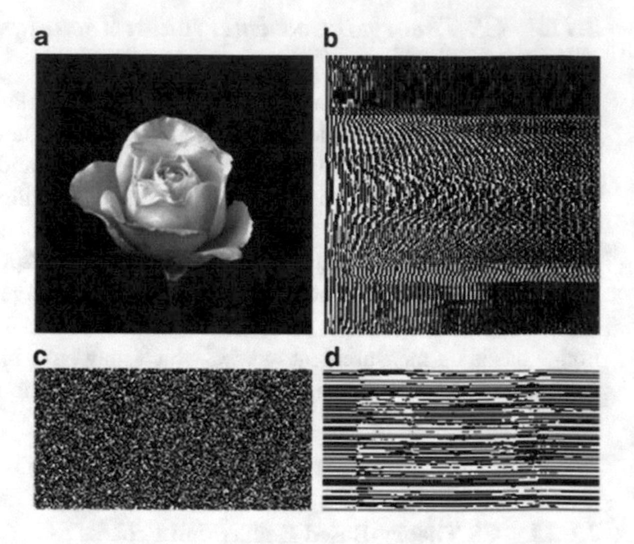

Fig. 2.4 Generation of encrypted image using CS theory-based encryption process. (**a**) Original image. (**b**) Sparse coefficients of image. (**c**) Measurement matrix. (**d**) Encrypted image (in term of CS measurements)

- Generate measurement matrix A with size of $M \times N$. Use A as secret key and decide a compression factor for the image.
- Generate the sparse measurements of the image by multiplying sparse coefficients with measurement matrix A.

$$\mathrm{Ef}_{M\times N} = A_{M\times N} \times x_{N\times N} \tag{2.7}$$

where Ef is compressed and encrypted image in terms of sparse measurements and A is the measurement matrix.

2.1.2.2 CS Theory-Based Decryption Process

In CS decryption process, the decryption of image takes place from its sparse measurement. The decryption process can be performed using CS recovery algorithms and measurement matrix. These recovery algorithms are based on linear algebra properties and optimization techniques. There are various types of CS recovery algorithms: L-norm minimization and greedy-based approach (Donoho 2006; Candes 2006; Baraniuk 2007; Tropp and Gilbert 2007; Nagesh and Li 2009). The CS theory-based decryption process for grayscale standard image is shown in Fig. 2.5.

Let Ef be the encrypted image after CS encryption; then the decryption of image Df can be performed using CS recovery algorithm (CSRA). The sparse coefficients of the image are in transform domain Ψ; the image is decrypted by using below equations:

$$x' = \mathrm{CSRA}(\mathrm{Ef}, A) \tag{2.8}$$

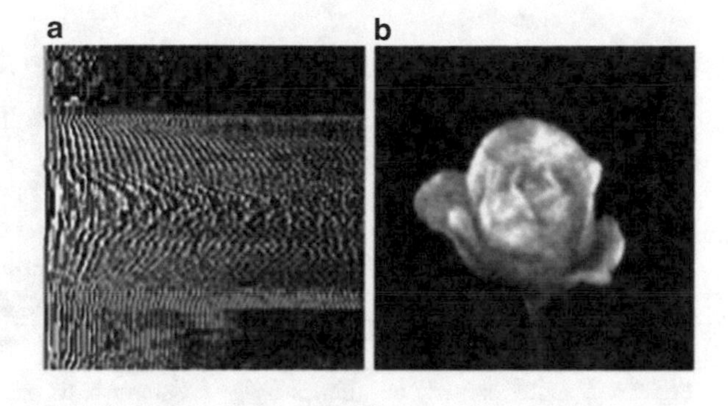

Fig. 2.5 Decrypted image using CS theory-based decryption process. (**a**) Decrypted Sparse coefficients of image. (**b**) Decrypted image

$$\mathrm{Df}_{N\times N} = \Psi'_{N\times N} \cdot x'_{N\times N} \cdot \Psi_{N\times N} \tag{2.9}$$

where x' is an extracted sparse coefficient of the image with size of $N \times N$, A is a measurement matrix with size of $N \times M$, Ef is an encrypted image with size of $M \times N$, and Df′ is a decrypted image with size of $N \times N$.

2.1.2.3 Various Important Properties of CS Theory

Some of the important properties of compressive sensing (CS) theory are given below. These properties are related to encryption process and decryption process.

- *Sparsity*

The image is defined as sparse if and only if when an image has few numbers of non-zero element (Donoho, 2006; Candes, 2006; Baraniuk, 2007). This property is related to CS theory acquisition procedure. The most natural and computerized images have a sparse representation when they are expressed into its transform domain.

When compressive sensing theory is applied to any image, the first step is that the image is converted into its sparse domain. The various transforms such as DCT, DWT, and SVD have sparsity property. When these transforms are applied to any image, the image is converted into its sparse domain. The sparse coefficients of the image are obtained by these image transforms. Figure 2.6 shows the sparsity property of different image transform basis matrices.

- *Incoherent Sampling*

This property is related to CS theory encryption process. When encryption of any image is performed using CS theory, the coherence of sensing image with respect to the transform matrix is given by Eq. 2.10:

$$\mu(A,\Psi) = \sqrt{N} \cdot \max_{1\le j\le m, 1\le j\le n}\left|\left\langle A_j, \Psi_i \right\rangle\right| \tag{2.10}$$

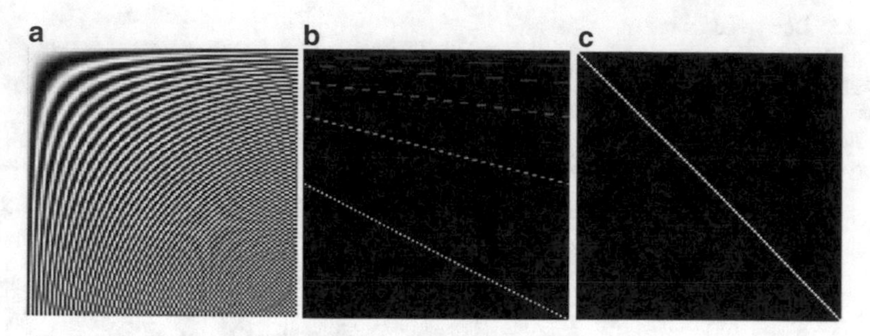

Fig. 2.6 Sparsity property of different image transform basis matrix. (**a**) DCT basis matrix. (**b**) DWT basis matrix. (**c**) SVD basis matrix

where Ψ and A are the transform basis and measurement matrix of R^N. From linear algebra, it follows that $\mu(A, \Psi) \in |1, \sqrt{N}|$.

If the measurement matrix and transform basis matrix are incoherent, then the image can be perfectly reconstructed from its sparse measurements.

- *Restricted Isometric Property (RIP)*

The necessary condition for an image decryption using CS measurements is that it has to satisfy restricted isometry property (RIP). The measurement matrix A with the size of $M{\times}N$ obeys the RIP of order K $(K < m)$ if measurement matrix A approximately preserves the squared magnitude of any K sparse encrypted image Ef using below equation.

$$\begin{aligned} &\forall y \text{ for which} \\ &\|y\|_0 \leq k, \\ &(1-\delta_K)\|\mathrm{Ef}\|^2 \leq \|A\mathrm{Ef}\|^2 \leq (1+\delta_K)\|\mathrm{Ef}\|^2 \\ &0 < \delta_k < 1 \end{aligned} \tag{2.11}$$

2.1.2.4 Information of Various CS Theory-Based Recovery Algorithms

There are various CS theory-based recovery algorithms described by researchers in the last 10 years for decryption of image from its CS measurements. There algorithms are based on linear algebra and convex optimization (Donoho 2006; Candes 2006; Baraniuk 2007; Candes and Romberg 2005; Tropp and Gilbert 2007; Gilbert et al. 2007; Laska et al. 2009; Dai and Milenkovic 2009; Needell 2009; Lopez and Boulgouris 2010; Duarte and Eldar 2011). These recovery algorithms are mainly divided into two types such as linear optimization-based (L_1 minimization technique) and greedy technique, respectively.

L_1 minimization technique (Donoho 2006; Candes 2006; Baraniuk 2007) is based on linear optimization. This algorithm provides good stability, but more computation time is required for decryption of image. When greedy-based CS recovery algorithms such as Orthogonal Matching Pursuit (OMP), Compressive Sampling Matching Pursuit (COSAMP), Greedy Basis Pursuit (GBP), Subspace Pursuit (SP), and Iterative Hard Thresholding (IHT) are based on iteration calculation of approximation of image coefficients. This algorithm is faster than minimization technique, but they do not provide stability (Tropp and Gilbert 2007; Gilbert et al. 2007; Laska et al. 2009; Dai and Milenkovic 2009; Needell 2009; Lopez and Boulgouris 2010; Duarte and Eldar 2011).

In this book, Orthogonal Matching Pursuit (OMP) algorithm is used for decryption of watermark biometric image at extraction side from its extracted CS measurements. The reason behind choosing this algorithm is that it is easy to understand and implement and required less computational time compared to the other CS recovery algorithms.

- *Orthogonal Matching Pursuit (OMP) Algorithm*

The Orthogonal Matching Pursuit (OMP) algorithm is introduced and designed by J. Tropp and A. Gilbert in 2007 (Tropp and Gilbert 2007). This algorithm is a greedy algorithm which is used for extraction of sparse coefficients from the CS measurements. The OMP algorithm is defined by three basic steps such as matching, orthogonal projection, and residual updating. The output of OMP algorithm is one non-zero sparse coefficient in each iteration. The OMP algorithm extracts sparse coefficients x from $y = Ax$. The mathematical steps for OMP algorithm are described in the following steps:

Inputs: CS measurements $\boldsymbol{y}$, measurement matrix $\boldsymbol{A}$
Output: Sparse coefficients $\boldsymbol{x}$
Initialization: Index $\boldsymbol{I} = \boldsymbol{A}$, residual $\boldsymbol{r} = \boldsymbol{y}$, sparse representation $\boldsymbol{\theta} = \boldsymbol{0} \in \boldsymbol{Rm}$

- *Step 1*: Initialize the residual $r_0 = y$ and the set of selected variables $x(c_0) = \phi$. Let iteration counter $i = 1$.

$$\max_t \left| x_t' r_{i-1} \right| \tag{2.12}$$

- *Step 2*: Find the variable x_t that solves the maximization problem given below using Eq. (2.4), and add the variable x_{ti} to the set of selected variables. Update C_i using Eq. (2.13).

$$C_i = C_{i-1} \cup \{t_i\} \tag{2.13}$$

- *Step 3*: Let P_i denote the projection onto the linear space spanned by the elements of $x(c_i)$. Then update residual r using Eq. (2.15).

$$P_i = x(C_i)\left(x(C_i)' x(C_i)\right)^{-1} x(C_i)' \tag{2.14}$$

$$r_i = (I - P_i)y \tag{2.15}$$

- *Step 4*: If the stopping condition is achieved, stop the algorithm. Otherwise, set $i = i + 1$ and return to step 2.

The solution of Eq. 2.14 is achieved by least squares optimization method. The value of projection P is taken as extracted sparse coefficients x. The value of extracted sparse coefficients depends on the linear projection between CS vector and measurement matrix-vector. Only when both the vectors have equal value, the output is zero because the projection is zero. Hence, the output of OMP algorithm is non-zero coefficients always.

2.2 Related Works

The literature review is presented in different sections such as works carried out in the area of biometric system and multibiometric system. The related works on biometric watermarking techniques and watermarking techniques with CS theory-based encryption are explained with its advantages and limitations. The basic algorithms and properties of CS theory are also covered in this section. Finally, the information about performance evaluation parameters for watermarking techniques is also covered in this section.

2.2.1 *Biometric System and Biometric Template Protection Technique Based on Cryptography*

The review papers based on various biometric traits, operation of biometric system, and the limitations related to biometric system are covered in this section. These papers are also given different security approaches using one-way transformation and cryptography for biometric template protection.

The report on biometric and its standards (Biometrics and Standards 2009) has explained various characteristics of different biometric templates. The report has also given the limitation of a unimodal biometric system and various evaluation parameters for performance of the biometric system. Another report on biometrics (Biometrics and Testing 2006) has explained various verification and authentication procedure for biometric recognition. A. Jain and A. Kumar (Jain and Kumar 2012) have explained various biometric traits of an individual. They have explained various issues such as designing of new biometric template protection technique against spoof attack and designing of new system for large-scale biometric data. They have also pointed out the need of new biometric recognition system for soft biometrics such as scars, marks, tattoos, color of the eye, and hair color. Jain and his research team (Jain et al. 2004) have explained various biometric recognition systems, errors that occur in the biometric system, comparison of various biometrics characteristics, and application of biometrics. They have also explained multibiometric system, social acceptance, and privacy issue of biometrics. A. Jain and K. Nandakumar (Jain and Nandakumar 2012; Nandakumar 2008) have explained operation of the biometric system. They have also pointed out the various attacks on the biometric system. They have designed various security techniques using cryptography for biometric template.

Nandakumar and his research team (Nandakumar 2008; Jain et al. 2008) have explained the advantages and disadvantages of various biometric template protection approaches such as bio-hashing technique, non-invertible transform technique, key binding-based biometric cryptosystem, and key generating-based biometric cryptosystem. They have also given some new research direction such as the need of developing hybrid biometric template protection technique. Jain and his research team (Jain et al. 2006) have explained the use of biometric data for security of multimedia information such as images and videos. They have also described issues regarding real-time implementation of biometric-based applications.

R. Giot and his research team (Giot et al. 2012) have given mathematical equations for evaluation parameters such as False Acceptance Rate (FAR), False Rejection Rate (FRR), and Equal Error Rate (EER) for biometric system as well as multibiometric system. N. Ratha and his research team (Ratha et al. 2001) have explained strengths of a biometric system. They have identified the weak links in the biometric system and give some solutions for these weak links. J. Pato and L. Millett (Pato and Millett 2010) have presented a report on the challenges and opportunities of a biometric system. In this report, they have given basic concepts of the biometric system, various operations, and evaluation parameters for the biometric system. They have also given future research opportunities of a biometric system.

H. Kolan and T. Thapaliya (Kolan and Thapaliya 2011) have presented a report on the application of biometrics for electronic passport security. They have also given an application of cryptography technique for biometric data security. Y. Sui and his research team (Sui et al. 2013) have explained the concept of cancellable biometrics for security and privacy of a biometric template. J. Ashbourn (Ashbourn 2014) has explained various biometric traits, usage of biometric traits, and its social application. He has also explained the application of biometrics in cloud computing and mobile phone. E. Harinda and E. Natgwirumugara (Harinda and Natgwirumugara 2015) have explained biometric-based identity card for Rwanda University. They have also explained challenges, parameters, and issues related to designing of biometric-based identity card.

2.2.2 Multibiometric System

The related works like operation, designing approaches, and application of the multibiometric system are covered in this section.

T. Panchal and A. Singh (Panchal and Singh 2013) have explained the advantages of the multibiometric system. They have also explained applications and future research direction in the multibiometric system. D. Gayathri and R. Rani (Gayathri and Rani 2013) have explained various technologies, different fusion procedures, designing issues, and various mode of operation for the multibiometric system. K. Sasidhar and his research team (Sasidhar et al. 2010) have explained limitations of unimodal biometric system and advantages of multibiometric system over unimodal biometric system. They have examined the accuracy and performance of a multibiometric system using commercial off-the-shelf (COTS) products.

M. Kaur and his research team (Kaur et al. 2010) have proposed a multibiometric system using signature and speech modalities to improved security and accuracy of multiple biometric data. E. Lupu and P. Pop (2008) have discussed various features of the multibiometric system such as its architecture, the level of fusion, and the methodologies used for identification of multiple biometric data. M. Agrawal (Agrawal 2007) has explained two design approaches for multistage identification of biometric data. These approaches are reduced time as well as improved accuracy of identification of biometric data. T. Ko (Ko 2005) has explained various fusion approaches for a multibiometric system based on the fingerprint, face, and iris. A. Ross and A. Jain (Ross and Jain 2004) have introduced a multibiometric system to overcome the limitations of unimodal biometric system. M. Theime (Theime 2003) has given various applications of the multibiometric system such as physical access, civil ID, and criminal ID. He also points out research challenges for designing of a multibiometric system.

2.2.3 Fragile Watermarking Techniques for Biometric Data Authentication

In this section, various watermarking techniques for biometric authentication are described. These techniques are implemented using various biometric traits such as the face, iris, and fingerprint. These techniques are used for authentication of biometric data against spoofing attack at system database of the biometric system.

V. Joshi (Joshi et al. 2013) has proposed a watermarking technique using multistage vector quantization and Discrete Cosine Transform (DCT) for fingerprint images. M. Joshi and his research team (Joshi et al. 2011) have proposed a Singular Value Decomposition (SVD)-based watermarking technique for fingerprint images. This proposed technique is used to provide authentication of biometric data at system database and matching subsystem. C. Li and his research team (Li et al. 2010) have introduced salient region-based watermarking for biometric template authentication. They have also proposed multilevel authentication of biometric data.

W. Kim and H. Lee (Kim and Lee 2009) have a blind spread spectrum-based watermarking technique for multiple biometric data authentication. This technique is used for two-stage secure identification of an individual. T. Hoang and his research team (Hoang et al. 2008) have proposed watermarking technique using a combination of amplitude modulation and bit priority level of feature values. This technique was designed for network-based biometric system. This proposed technique has improved the security as well as reduces the bandwidth for biometric data. J. Picard (Picard et al. 2004) has explained various techniques for security enhancement of identification cards, passports, and licenses. He has explained digital watermarking, 2D-based codes, and copy detection pattern for biometric data authentication. S. Pankanti and M. Yeung (Pankanti and Yeung 1999) have proposed WSQ-based watermarking technique. In this technique, the identification of an individual was performed by watermarked fingerprint image. This technique is used for copyright authentication applications.

2.2.4 Robust Watermarking Techniques for Biometric Data Protection

In this section, various watermarking techniques for biometric protection are described. These techniques are implemented using various biometric traits such as the face, iris, and fingerprint. These techniques are used for protection of biometric data against modification attack at the communication channel between two modules of the biometric system.

V. Inamdar and P. Rege (Inamdar and Rege 2014) have proposed a dual watermarking technique using DCT and Discrete Wavelet Transform (DWT) for multiple biometric data. In this technique, the compressed speech signal coefficients and Gabor facial features were taken as watermark information, and this watermark information are embedding various transform coefficients of standard image. This technique provides multiple-level ownership for multimedia data using biometrics. N. Chaudhary (Chaudhary et al. 2013) has proposed a DWT-, DCT-, and SVD-based watermarking technique. In this technique, wavelet coefficients of watermark fingerprint image are inserted into the singular value of hybrid coefficients of host face image. The hybrid coefficients of host face image are generated by DWT and DCT. A. Javed (Javed et al. 2013) has proposed an additive watermarking technique for protection of face image and fingerprint image. The robustness of this technique is enhanced by using repetitive coding. B. Behera and V. Govindan (Behera and Govindan 2013) have proposed DCT and phase congruency model-based watermarking technique for multiple biometric data. This technique is used in e-passport and e-identification cards.

P. Tamije Selvy (Tamije Selvy et al. 2013) has proposed a watermarking technique using wavelet-based contour transform. In this technique, a color watermark iris data are inserted into the second-level wavelet-based contourlet transform coefficients of color standard image. M. Paunwala and S. Patnaik (Paunwala and Patnaik 2014) have proposed a watermarking technique using DCT. In this technique, fused multiple biometric data are inserted into low-frequency AC coefficients of 8 × 8 DCT blocks of the standard image. The fused multiple biometric data are generated using fingerprint features and iris features of an individual. R. Rege (Rege 2012) has proposed a correlation-based watermarking technique for protection of offline handwritten signature image. She has also explained limitations of biometric watermarking, and less standard algorithms are available for biometric data protection.

V. Inamdar and P. Rege (Inamdar and Rege 2012) have proposed a semi-blind watermarking technique for fingerprinting application. They have first extracted facial features using principal component analysis (PCA). These features are inserted into singular values of the standard host image. M. Isa and S. Aljareh (Isa and Aljareh 2012) have described combination of watermarking and face recognition for biometric image protection. They have used Cox watermarking algorithm using DCT and embedded secure password as a watermark into host face image. They have also explained that the performance of face recognition rate does not change due to watermarking. P. Bedi (Bedi et al. 2012; Bedi et al. 2011) have proposed a DCT and particle swarm optimization (PSO)-based watermarking technique for security of biometric data. In this technique, PSO algorithm is used to find best DCT coefficients of host face image for watermark embedding. The watermark data is inserted into these DCT coefficients of host face image using quantization-based watermarking technique.

C. Li (Li et al. 2011) has proposed a blind SS-QIM-based watermarking technique for enhancement of security of multiple biometric data. In this technique, first, facial features are converted into a raw data which are taken as watermark data. These watermark data is inserted into a host fingerprint image. S. Edward (Edward et al. 2011) has proposed a new authentication technique using multiple biometric verifications and watermarking for digital rights management (DRM). In this technique, face image is taken as host, and iris image is taken as watermark. This technique used ridgelet transform, and these transform is applied to watermark iris image to its ridgelet transform coefficients. These coefficients are inserted into a host face image to get secure face image.

V. Inamdar (Inamdar et al. 2010) has proposed a watermarking technique using biorthogonal wavelet transform for handwritten signature protection. They also explained signature template matching procedure for the identification of an individual. V. Jundale and S. Patil (Jundale and Patil 2010) have proposed a watermarking technique using wavelet transform for speech signal protection. In this technique, speech signal is taken as watermark and inserted into a standard image. D. Mathivadhani and C. Meena (Mathivadhani and Meena 2010) have given a study on various watermarking technique using DCT and DWT for fingerprint image protection. R. Motwani (Motwani 2010) has proposed a watermarking technique for protection of the voice of a person. She has explained that the voiceprint of a person is created by using Mel-frequency Cepstral coefficients of the speech signal and Gaussian mixture model. B. Ma (Ma et al. 2010) has proposed a QIM-based watermarking technique for multiple biometric data protection. In this technique, fingerprint minutiae points are embedded into a block pyramid layer of the region of the host face image.

Y. Cao (Cao et al. 2010) has proposed a watermarking technique using contourlet transform and quantization. In this technique, watermark biometric data are inserted into contourlet coefficients of host biometric data. Qi (Qi et al. 2010) have proposed a correlation-based watermarking technique for security and privacy of biometric data. In this technique, it has first found a correlation between watermark biometric image and host image using partial least squares (PLS) and particle swarm optimization (PSO). Based on this correlation data analysis, watermark biometric data are inserted into the host image. A. Naik and R. Holambe (Naik and Holambe 2010) have proposed watermarking technique using mapping. In this technique, watermark fingerprint image is inserted into selected DCT coefficients of host face image. K. Zebbiche (Zebbiche et al. 2008) has proposed region of interest (ROI)-based watermarking technique. In this technique, watermark data is inserted into the ROI of the fingerprint image.

A. Noore (Noore et al. 2007a, b) has proposed a watermarking technique using Discrete Wavelet Transform (DWT). In this technique, selected face image and corresponding texture information are inserted into the selected texture region of the fingerprint image. They have also claimed that this technique has improved security of an automatic fingerprint identification system. M. Vatsa (Vatsa et al. 2009) has proposed watermarking technique using three-level redundant DWT and phase congruency model. K. Park (Park et al. 2007) has described transform domain watermarking technique for multiple data protection. This technique was designed for protection of iris features and facial features. M. Vatsa (Vatsa et al. 2005) has described a watermarking technique using DWT and Least Significant Bit (LSB) substitution. In this technique, LSB of facial features are inserted into wavelet coefficients of fingerprint image. D. Moon (Moon et al. 2005) has described different watermarking techniques for protection of multiple biometric data. In this technique, first, facial features are inserted into the fingerprint features. These fingerprint features are inserted into the host image. The authors also claimed that this technique has improved the accuracy of identification of an individual.

M. Vatsa (Vatsa et al. 2005) has described watermarking technique using multi-resolution DWT and Support Vector Machine (SVM). The authors claimed that face recognition process is improved by 10% for watermarked face image under various attacks by using ν-Support Vector Machine-based intelligent algorithm. A. Giannoula and D. Hatzinakos (Giannoula and Hatzinakos 2004) have described a watermarking technique using DWT for security of iris and speech signal. In this technique, speech signal coefficients and iris pattern are inserted into wavelet coefficients of fingerprint image using block process. The authors applied compression to a watermarked multibiometric data to improve storage capacity of system database. M. Vatsa (Vatsa et al. 2004) has proposed two watermarking techniques using correction property of PN sequences and DCT. These techniques provided two levels of verification where watermarked face image is used for first-level verification and extracted iris image is used for cross verification of an individual.

2.2.5 *Watermarking Techniques with Compressive Sensing (CS) Theory Procedure*

In this section, various watermarking techniques with compressive sensing (CS) theory procedure are described. These techniques used CS theory procedure as encryption and decryption of watermark data before embedding into the host data. These techniques are used for authentication and tamper identification for various types of data.

A. Sreedhanya and K. Soman (Sreedhanya and Soman) have proposed a watermarking technique using SVD and CS theory procedure for security of medical image for telemedicine application. In this technique, watermarked medical image is encrypted into sparse data using CS theory procedure. This encrypted medical image is inserted into the singular values of the host image. F. Tiesheng (Tiesheng et al. 2013) has proposed a watermarking technique using CS theory procedure for multimedia data authentication. In this technique, at embedder side, CS theory-based encryption is applied to watermark data for get encrypted watermark data. At extractor side, CS theory-based decryption is applied to extracted encrypted watermark data to get actual watermark data. M. Fakhr (Fakhr 2012) has proposed a watermarking technique using CS theory procedure for audio signal. He has also given a comparison of the effect of MP3 audio compression and additive noise on watermarked audio signal.

M. Raval (Raval et al. 2011) has proposed a watermarking technique using DWT and CS theory procedure for tamper detection in the image. X. Zhang (Zhang et al. 2011) has proposed a watermarking technique using flexible self-recovery quality-based CS theory and DCT. This technique was proposed for image tampering identification. M. Tagliasacchi (Tagliasacchi et al. 2009) has proposed a watermarking technique for image tampering identification. M. Sheikh and R. Baraniuk (Sheikh and Baraniuk 2007) have introduced the application of CS theory procedure in the digital watermarking area. The authors have combined CS theory procedure with spread spectrum watermarking technique for multimedia data protection.

2.2.6 *Compressive Sensing (CS) Theory, Its Reconstruction Algorithms, and Application of CS Theory in Information Security Field*

In this section, basic reference papers related to CS theory are given. These papers have given information for CS theory, its properties, mathematical formula, and various reconstruction algorithms. This section provides also application of CS theory in the information security field. The use of CS theory in this field has started around 2013, and various approaches are designed and implemented by researchers.

D. Donoho (Donoho 2006) has introduced a new signal acquisition theory which is called as compressive sensing or sampling (CS). He has also described that it is possible to reconstruct signals or images accurately and exactly from number of few samples which are far smaller than the desired resolution of the signal or image, e.g., pixels of the image. E. Candes (Candes 2006) has given mathematical models, properties, and various theorems for CS theory. R. Baraniuk (Baraniuk 2007) has given a summary of CS theory. He has also given an overview of compressive sensing reconstruction algorithm which is based on L-norm minimization. E. Candes and J. Romberg (Candes and Romberg 2005) have described convex optimization programming and L_1 minimization technique for image reconstruction from inaccurate sampled and corrupted measurement. J. Tropp and A. Gilbert (Tropp and Gilbert 2007) have given the theory of Orthogonal Matching Pursuit (OMP). They have explained that this algorithm is reliable to recover a signal or image from its sparse measurement. This algorithm is faster and easier to implement compared to the other CS reconstruction algorithms. A. Gilbert (Gilbert et al. 2007) has described the properties, rules, and required computation time for CS reconstruction algorithms. He has also proposed a new CS reconstruction algorithm such as Heavy Hitters on Steroids (HHS) pursuit and compared this algorithm with other CS reconstruction algorithms.

J. Laska (Laska et al. 2009) has described the pursuit of justice and basis pursuit denoising algorithm for signal or image reconstruction from sparsely corrupted measurements. W. Dai and O. Milenkovic (Dai and Milenkovic 2009) have proposed subspace pursuit algorithm for reconstruction of sparse signals with and without noise. This algorithm has important characteristics such as low computational complexity and applied to very sparse signal compared to other greedy CS reconstruction algorithms. D. Needell (Needell 2009) has given mathematical derivations and MATLAB implement of different greedy algorithms. R. Lopez and N. Boulgouris (Lopez and Boulgouris 2010) have described compressive sensing procedure for image compression application. They have also compared this proposed compression standard with existing image compression standards. M. Duarte and Y. Eldar (Duarte and Eldar 2011) have described a complete summary about the compressive sensing theory procedure and CS reconstruction algorithms such as L-norm minimization and greedy algorithms for signal reconstruction.

After the invention of CS theory around 2006, researchers introduced this theory in various applications such as image security (image encryption, image watermarking, image hashing, image hiding, and authentication), video security, cloud security scenario, and 5G system security scenario (Zhang et al. 2016). In the last 8 years, researchers designed approaches using CS theory process with some cryptographic-based encryption techniques applied on images. Table 2.1 shows the summary of CS-based approaches used in combination with cryptographic-based encryption techniques to provide security to images.

Table 2.1 Summary of CS-based procedures with combination of cryptographic-based encryption technique for security of images

Sl. No.	Existing techniques	Type of compression method	Type of encryption method	Nature of key in encryption method
1	Zhou et al. (2014)	CS	Random pixel scrambling	Symmetric
2	Huang et al. (2014)	CS	Arnold scrambling, block-wise XOR operation	Symmetric
3	Fira (2015)	CS	Substitutions	Symmetric
4	Zhang et al. (2015)	Random convolution and subsampling method-based CS	Linear transform based encryption	Symmetric
5	Ahmed et al. (2016)	DCT, orthogonal matrix	Partial encryption	Symmetric
6	Chen et al. (2016)	Kronecker CS	Elementary cellular automata (ECA) scrambling	Symmetric
7	Deng et al. (2017)	2D Cs	Discrete fractional random transform (DFrRT)	Symmetric
8	Zhou et al. (2016)	2D Cs	Cycle shift operation	Symmetric

2.2.7 Multimedia Watermarking Techniques

In this section, basic theory, properties, and general framework for multimedia watermarking techniques in the spatial and transform domain are given.

A. Kothari (Kothari 2013) has proposed and compared various watermarking techniques using DCT, DWT, and SVD for copyright protection of digital video. R. Jahan (Jahan 2013) has proposed watermarking technique using DWT-SVD and optimized chaotic-based encryption method. In this technique, first, watermark is encrypted using genetic algorithm and chaotic function. This encrypted watermark data is inserted into singular values of wavelet coefficients of the host image to get watermarked image. Bazargani (Bazargani et al. 2012) has given a comparison of watermarking techniques in wavelet, contourlet, and curvelet for standard image protection. A. Gupta and M. Raval (Gupta and Raval 2012) have proposed watermarking techniques using pure SVD and DWT + SVD for copyright protection. R. Thanki (Thanki et al. 2011) has described and given a comparison of various visible and invisible watermarking techniques for copyright protection of grayscale images and color images using WGN and PN sequences in the spatial domain.

H. Tsai and C. Liu (Tsai and Liu 2011) have proposed a wavelet-based watermarking technique for copyright protection using the HVS model and neural networks. In this technique, just noticeable difference (JND) profile of watermark image is embedded into wavelet coefficients of the host image. They have also described that an artificial neural network is used in the proposed technique for memorizing the difference between original wavelet coefficients and its watermarked version. D. Shinfeng (Shinfeng et al. 2010) has proposed a DCT-based image watermarking technique. This technique has improved robustness against JPEG compression. In this technique, low-frequency DCT coefficients of host image are replaced by watermark data without introducing degradation in the quality of watermarked image. J. Xu (Xu et al. 2010) has proposed a watermarking technique based on Fast Discrete Curvelet Transform for standard image protection. In this technique, the binary watermark data is embedded into low-frequency curvelet coefficients of the host image. M. Rohani and A. Avanaki (Rohani and Avanaki 2009) have proposed a PSO- and DCT-based image watermarking technique. In this technique, particle swarm optimization (PSO) is used to find the best DCT coefficients of host image where watermark sequences are inserted.

A. Mansouri (Mansouri et al. 2009) has proposed a non-blind image watermarking technique. In this technique, singular values of complex wavelet coefficients of host image are modified according to the value of the watermark image. This proposed technique is robust against most common watermarking attacks. S. Hajjara (Hajjara et al. 2009) has proposed a biorthogonal wavelet transform-based robust watermarking technique for copyright protection. In this technique, watermark data is embedded into horizontal wavelet coefficients of the host image. G. Bhatnagar and B. Raman (Bhatnagar and Raman 2009) have described SVD- and DWT-based robust watermarking technique for copyright protection. In this technique, singular values of wavelet coefficients of host image are modified according to singular values of watermark image. C. Zhang (Zhang et al. 2008) has proposed a first watermarking technique based on curvelet transform for copyright protection. In this technique, image hash is computed using curvelet transform. These hash values are embedded into a different level of curvelet transform of the host image.

R. Dili and E. Mwangi (Dili and Mwangi 2007) have proposed SVD-based robust watermarking technique. In this technique, the monochrome watermark image is embedded into singular values of selected wavelet blocks of horizontal and vertical subbands of the host image. The embedded blocks are selected by a secret key to enhance imperceptibility of technique. E. Ganic and M. Eskicioglu (Ganic and Eskicioglu 2004) have proposed an image watermarking technique for copyright authentication. In this technique, quantization values of singular values of host image blocks are modified according to watermark data in the spatial domain. C. Chan and L. Cheng (Chan and Cheng 2004) have described the simple data hiding technique based on LSB substitution for image protection.

M. Raval and P. Rege (Raval and Rege 2003) have proposed wavelets and SVD-based adaptive watermarking technique. This technique utilized singular values of the blocks of a wavelet subband of the host image. They have claimed that if the watermark is embedded into low-frequency coefficients, then the technique is robust against LPF, JPEG compression, and geometric attacks. If a watermark is embedded into higher-frequency coefficients, then the technique is robust against contrast, histogram equalization, and cropping attacks. G. Langelaar (Langelaar et al. 2000) has given a review on properties, applications, techniques, and requirements of watermarking. They have beautifully explained what is actually happening when embedding watermark image in the cover image using watermarking in the spatial domain and transform domain. They have also explained the watermarking technique for video data. I. Cox (Cox et al. 1997) has described a robust watermarking technique based on correlation properties of PN sequence and spread spectrum technique. They have given standard watermarking methodology which is used for any multimedia data.

2.2.8 *Evaluation Parameters for Watermarking Techniques and Biometric System*

In this section, information about evaluation parameters for watermarking techniques and biometric system is given.

In the various papers (Wang and Bovik, 2004; Voloshynovskiy et al., 2001; Petitcolas, 2000; Wolfgang et al., 1999; Kutter and Petitcolas, 1999), authors have described mathematical expressions of visual quality matrices for performance evaluation of watermarking techniques. They provided common evaluation parameters which are used by the inventor for checking performance of proposed technique with previously existing techniques. They are also explained various attacks that degraded the quality of watermarked data.

In the various papers (Biometrics & Standards, 2009; Giot et al., 2012; Fernandez et al., 2012; Behaviometics, 2009), authors have described evaluation parameters for the performance of the biometric system. They have also given mathematical expressions of evaluation parameters such as the False Rejection Rate (FRR), False Acceptance Rate (FAR), and Equal Error Rate (EER).

2.2.9 *Biometric Recognition Techniques*

In this section, information related to biometric recognition techniques used in the presented work are given.

A. Jain and S. Prabhakar (Prabhakar 2001; Jain et al. 1999) have described fingerprint recognition algorithm based on a filter bank classification. This algorithm has given the Euclidean distance between query fingerprint image and enrolled fingerprint image. Based on the result of template matching, human recognition has taken place. In the review papers (Yang et al., 2000; Lu et al., 2003), authors have described face recognition algorithm based on linear discriminator analysis (LDA) (Welling 2005). This algorithm has given the Euclidean distance between query face image and enrolled face image. Based on results of template matching, human recognition has taken place.

References

Agrawal, M. (2007). *Design approaches for multimodal biometric system*. M. Tech. thesis, Department of Computer Science and Engineering, IIT, Kanpur.

Ahmed, J., Khan, M., Hwang, S., & Khan, J. (2016). A compression sensing and noise-tolerant image encryption scheme based on chaotic maps and orthogonal matrices. *Neural Computing and Applications, 2016*, 1–5.

Ashbourn, J. (2014). *Biometrics in the new world: The cloud, mobile technology and pervasive identity*. Cham: Springer.

Baraniuk, R. (2007). Compressive sensing. *IEEE Signal Processing Magazine, 24*, 118–124.

Bazargani, M., Ebrahimi, H., & Dianat, R. (2012). Digital image watermarking in wavelet, contourlet and curvelet domains. *Journal of Basic and Applied Scientific Research, 2*(11), 11296–11308.

Bedi, P., Bansal, R., & Sehgal, P. (2011). Using PSO in image hiding scheme based on LSB substitution. *Advances in Computing and Communications*, 2011: 259–268.

Bedi, P., Bansal, R., & Sehgal, P. (2012). Multimodal biometric authentication using PSO based watermarking. *Procedia Technology, 4*, 612–618.

Behaviometrics. (2009). *Measuring FAR/FRR/EER in continuous authentication*. A Technical White Paper, BehavioSec.

Behera, B., & Govindan, V. (2013). Improved multimodal biometric watermarking in authentication systems based on DCT and phase congruency model. *International Journal of Computer Science and Network, 2*(3), 123–129.

Bhatnagar, G., & Raman, B. (2009). A new robust reference watermarking scheme based on DWT-SVD. *Computer Standards & Interfaces, 31*, 1002–1013.

Biometrics and Standards. (2009). *ITU-T Technology Watch Report, December*. Available http://www.itu.int/dms_pub/itu-t/oth/23/01/T230100000D0002MSWE.doc

Biometrics Testing and Statistics. (2006). *National Science and Technology Council (NSTC) Report*. Available www.biometrics.gov/documents/biotestingandstats.pdf

Candes, E. (2006). Compressive sampling. *Proceedings of the International Congress of Mathematicians*. pp. 1–20.

Candes, E., & Romberg, J. (2005). L1-Magic: Recovery of Sparse signals via convex programming. pp. 1–19.

Cao, Y., Gong, W., Cao, M., & Bai, S. (2010). Robust biometric watermarking based on contourlet transform for fingerprint and face protection. *Proceedings of 2010 I.E. International Symposium on Intelligent Signal Processing and Communication Systems (ISPACS)*, pp. 1–4.

Chan, C., & Cheng, L. (2004). Hiding data in images by simple LSB substitution. *Pattern Recognition, 37*, 469–474.

Chaudhary, N., Singh, D., & Hussain, D. (2013). Enhancing security of multimodal biometric authentication system by implementing watermarking utilizing DWT and DCT. *IOSR Journal of Computer Engineering, 15*(1), 6–11.

Chen, T., Zhang, M., Wu, J., Yuen, C., & Tong, Y. (2016). Image encryption and compression based on kronecker compressed sensing and elementary cellular automata scrambling. *Optics & Laser Technology, 84*, 118–133.

Cox, I., Kilian, J., Shamoon, T., & Leighton, F. (1997). Secure spread spectrum watermarking for multimedia. *IEEE Transactions on Image Processing, 6*(12), 1673–1687.

Dai, W., & Milenkovic, O. (2009). Subspace pursuit for compressive sensing signal reconstruction. *IEEE Transactions on Information Theory, 55*(5), 2230–2249.

Deng, J., Zhao, S., Wang, Y., Wang, L., Wang, H., & Sha, H. (2017). Image compression – Encryption scheme combining 2D compressive sensing with discrete fractional random transform. *Multimedia Tools and Applications, 76*(7), 10097–10117.

Dili, R., & Mwangi, E. (2007). An image watermarking method based on the singular value transformation and the wavelet transformation. *Proceedings of IEEE AFRICON*, pp. 1–5.

Donoho, D. (2006). Compressed sensing. *IEEE Transaction on Information Theory, 52*(4), 1289–1306.

Duarte, M., & Eldar, Y. (2011). Structured compressed sensing: From theory to applications. *IEEE Transactions on Signal Processing, 59*(9), 4053–4085.

Edward, S., Sumanthi, S., & Ranihemamalini, R. (2011). Person authentication using multimodal biometrics with watermarking. *Proceedings of 2011 International Conference on Signal Processing, Communication, Computing and Networking Technologies (ICSCCN)*, pp. 100–104.

Fakhr, M. (2012). Robust watermarking using compressed sensing framework with application to MP3 audio. *The International Journal of Multimedia & Its Applications (IJMA), 4*(6), 27–43.

Fernandez, F., Fierrez, J., & Garcia, J. (2012). Quality measures in biometric systems. *IEEE Security and Privacy*, 52–62.

Fira, M. (2015). *Applications of compressed sensing: Compression and encryption*. 2015 E-Health and Bioengineering Conference (EHB): 1–4.

Ganic, E. and Eskicioglu, A. (2004). *Secure DWT-SVD Domain image watermarking: Embedding data in all frequencies*. ACM Multimedia and Security Workshop 2004, Magdeburg, Germany, pp. 1–9.

Gayathri, D., & Rani, R. (2013). Multimodal biometric system: An overview. *International Journal of Advanced Research in Computer and Communication Engineering, 2*(1), 898–902.

Giannoula, A. and Hatzinakos, D. (2004). Data hiding for multimodal biometric recognition. *Proceedings of the 2004 I.E. International Symposium on Circuits and Systems 2*, 160–165.

Gilbert, A., Strauss, M., Tropp, J., & Vershynin, R. (2007). One sketch for all: Fast algorithms for compressed sensing. *39th ACM Symposium on Theory of Computing (STOC)*. pp. 237–246.

Giot, R., El-Abed, M., & Rosenberger, C. (2012). Fast computation of the performance evaluation of biometric systems: Application to multibiometrics. *Future Generation Computer Systems, 1*, 1–30.

Gonzales, R., & Woods, R. (2002). *Digital image processing* (pp. 222–226). Upper Saddle River: Prentice Hall, Inc..

Gupta, A., & Raval, M. (2012). A robust and secure watermarking scheme based on singular value replacement. *Sadhana © Indian Academy of Science, 37*(4), 425–440.

Hajjara, S., Abdallah, M., & Hudaib, A. (2009). Digital image watermarking using localized biorthogonal wavelets. *European Journal of Scientific Research, 26*(4), 594–608.

Harinda, E., & Natgwirumugara, E. (2015). Security & privacy implications in the placement of biometric-based ID card for Rwanda Universities. *Journal of Information Security, 6*, 93–100.

Hoang, T., Dat, T., & Sharma, D. (2008). Remote multimodal biometric authentication using bit priority-based fragile watermarking. *Proceedings of 19th IEEE International Conference on Pattern Recognition (ICPR 2008)*, pp. 1–4.

Huang, R., Rhee, K., & Uchida, S. (2014). A parallel image encryption method based on compressive sensing. *Multimedia Tools and Applications, 72*(1), 71–93.

Inamdar, V., & Rege, P. (2012). Face features based biometric watermarking of digital image using singular value decomposition for fingerprinting. *International Journal of Security and Its Applications, 6*(2), 47–60.

Inamdar, V., & Rege, P. (2014). Dual watermarking technique with multiple biometric watermarks. *Sadhana © Indian Academy of Science, 29*(1), 3–26.

Inamdar, V., Rege, P., & Arya, M. (2010). Offline handwritten signature based blind biometric watermarking and authentication technique using biorthogonal wavelet transform. *International Journal of Computer Applications, 11*(1), 19–27.

Isa, M., & Aljareh, S. (2012). Biometric image protection based on discrete cosine transform watermarking technique. *Proceeding of International Conference on Engineering and Technology (ICET)*, pp. 1–5.

Jahan, R. (2013). Efficient and secure digital image watermarking scheme using DWT-SVD and optimized genetic algorithm based chaotic encryption. *International Journal of Science, Engineering and Technology Research (IJSETR), 2*(10), 1943–1946.

Jain, A., & Kumar, A. (2012). Biometric recognition: An overview. In E. Mordini & D. Tzovaras (Eds.), *Second Generation Biometrics: The Ethical, Legal and Social Context* (pp. 49–79). Dordrecht: Springer.

Jain, A., & Nandakumar, K. (2012). *Biometric authentication: System security and user privacy. IEEE Computer Society, 45*, 87–92.

Jain, A., Prabhakar, S., & Pankanti, S. (1999). *A Filterbank based representation for classification and matching of fingerprint*. International Joint Conference on Neural Networks (IJCNN), Washington, DC, July, pp. 3284–3285.

Jain, A., Ross, A., & Prabhakar, S. (2004). An introduction to biometric recognition. *IEEE Transactions on Circuits and Systems for Video Technology, Special Issue on Image and Video Based Biometrics, 14*(1), 4–20.

Jain, A., Ross, A., & Pankanti, S. (2006). Biometrics: A tool for information security. *IEEE Transactions on Information Forensics and Security, 1*(2), 125–143.

Jain, A., Nandakumar, K., & Nagar, A. (2008). Biometric template security. *EURASIP Journal on Advances in Signal Processing, Special Issue on Advanced Signal Processing and Pattern Recognition Methods for Biometrics*, January, pp. 1–17.

Javed, A., Fasihullah, M., Munir, M., Usman, I., Shafique, M., Bashir, T., & Khan, M. (2013). A new additive watermarking technique for multimodal biometric identification. *Journal of Basic and Applied Scientific Research, 3*(7), 935–942.

Joshi, M., Joshi, V., & Raval, M. (2011). Multilevel semi-fragile watermarking technique for improving biometric fingerprint system security. In A. Agrawal, R. C. Tripathi, E. Y.-L. Do, & M. D. Tiwari (Eds.), *Intelligent interactive technologies and multimedia* (pp. 272–283). Berlin/Heidelberg: Springer.

Joshi, V., Raval, M., Rege, P., & Parulkar, S. (2013). Multistage VQ based exact authentication for biometric images. *Computer Society of India (CSI) Journal of Computing, 2*(1–2), R3-25–R3-29.

Jundale, V., & Patil, S. (2010). Biometric speech watermarking technique in images using wavelet transform. *IOSR Journal of Electronics and Communication Engineering (IOSR-JECE)*, pp. 33–39.

Kaur, M., Girdhar, A., & Kaur, M. (2010). Multimodal biometric system using speech and signature modalities. *International Journal of Computer Applications, 5*(12), 13–16.

Kim, W., & Lee, H. (2009). Multimodal biometric image watermarking using two stage integrity verification. *Signal Process, 89*(12), 2385–2399.

Ko, T. (2005). Multimodal biometric identification for large user population using fingerprint, face and iris recognition. *Proceeding of the IEEE 34th Applied Imagery and Pattern Recognition Workshop (AIPR05).*

Kolan, H., & Thapaliya, T. (2011). *Biometric passport: Security and privacy aspects of machine readable travel document*. Available https://diuf.unifr.ch/main/is/sites/diuf.unifr.ch.main.is/files/documents/student-projects/eGov_2011_Hesam_Kolahan_&_Tejendra_Thapaliya.pdf

Kothari, A. (2013). *Design, implementation and performance analysis of digital watermarking for video*. Ph.D. thesis, JJTU, India.

Kutter, M., & Petitcolas, F. (1999). A fair benchmark for image watermarking systems. Electronic Imaging' 99. *Security and Watermarking of Multimedia Contents, 3657*, 1–14.

Langelaar, G., Setyawan, I., & Lagnedijk, R. (2000). Watermarking of digital image and video data – A state of art review. *IEEE Signal Processing Magazine*, 20–46.

Laska, J., Davenport, M., & Baraniuk, R. (2009). *Exact signal recovery from sparsely corrupted measurements through the pursuit of justice*. Asilomar Conference on Signals, Systems and Computers, pp. 1556–1560.

Li, C., Ma, B., Wang, Y., & Zhang, Z. (2010). *Protecting biometric templates using authentication watermarking, PCM 2010, Part I, LNCS 6297* (pp. 709–718). Berlin/Heidelberg: Springer.

Li, C., Ma, B., Wang, Y., & Zhang, Z. (2011). Sparse reconstruction based watermarking for secure biometric authentication. *Biometric Recognition, 7908*, 244–251.

Lopez, R., & Boulgouris, N. (2010). *Compressive sensing and combinatorial algorithms for image compression. A project report*. London: King's College London.

Lu, J., Plataniotis, N., & Venetsanopoulos, A. (2003). Face recognition using LDA based algorithms. *IEEE Transactions on Neural Networks, 14*(1), 195–200.

Lupu, E., & Pop, P. (2008). Multimodal biometric systems overview. *ACTA Technica Napocensis, 49*(3), 39–44.

Ma, B., Li, C., Wang, Y., & Zhang, Z. (2010). Block pyramid based adaptive quantization watermarking for multimodal biometric authentication. *Proceedings of 20th IEEE International Conference on Pattern Recognition (ICPR)*, pp. 1277–1280.

Mansouri, A., Aznaveh, A., & Azar, F. (2009). SVD based digital image watermarking using complex wavelet transform. *Sadhana © Indian Academy of Science, 34*(3), 393–406.

Mathivadhani, D., & Meena, C. (2010). A comparative study of fingerprint protection using watermarking techniques. *Global Journal of Computer Science and Technology, 9*(5), 98–102.

Moon, D., Tachae, K., Jung, S., Chung, Y., Moon, K., Ahn, D., & Kim, S. (2005). Performance evaluation of watermarking techniques for secure multimodal biometric systems. *Computational Intelligence and Security*, pp. 635–642.

Motwani, R. (2010). *A Voice-based biometric watermarking scheme for digital rights management of 3D mesh models*. Ph.D. thesis, University of Nevada, Reno.

Nagesh, P., Li, B. (2009). *Compressive imaging of color images*. 2009 I.E. International Conference on Acoustics, Speech and Signal Processing, pp. 1261 – 1264.

Naik, A., & Holambe, R. (2010). Blind DCT Domain digital watermarking for biometric authentication. *International Journal of Computer Applications (IJCA), 16*(1), 11–15.

Nandakumar, K. (2008). *Multibiometric systems: Fusion strategies and template security*. Ph.D. thesis, Michigan State University, USA.

Needell, D. (2009). *Topics in compressed sensing*. Ph.D. thesis, University of California, USA.

Noore, A., Singh, R., Vatsa, M., & Houck, M. (2007a). Enhancing security of fingerprints through contextual biometric watermarking. *Forensic Science International, 169*(2), 188–194.

Noore, A., Singh, R., Vatsa, M., Houck, M., & Morris, K. (2007b). Robust biometric image watermarking for fingerprint and face template protection. *IEICE Electronics Express, 3*(2), 23–28.

Panchal, T., & Singh, A. (2013). Multimodal biometric system. *International Journal of Advanced Research in Computer Science and Software Engineering, 3*(5), 1360–1363.

Pankanti, S. and Yeung, M. (1999). Verification watermarks on fingerprint recognition and retrieval. In *Electronic Imaging '99, International Society for Optics and Photonics*, pp. 66–78.

Park, K., Jeong, D., Kang, B. and Lee E. (2007). A study on iris feature watermarking on face data. *Adaptive and natural computing algorithms*, pp. 415–423.

Pato, J., & Millett, L. (2010). *Biometric recognition: Challenges and opportunities*. Whither Biometric Board. Available http://dataprivacylab.org/TIP/2011sept/Biometric.pdf

Paunwala, M., & Patnaik, S. (2014). Biometric template protection with DCT based watermarking. *Machine Vision and Applications, 25*(1), 263–275.

Petitcolas, F. (2000). Watermarking schemes evaluation. *IEEE Signal Processing Magazine, 17*, 58–64.

Picard, J., Vielhauer, C., & Thorwirth, N. (2004). Towards fraud-proof ID documents using multiple data hiding technologies and biometrics. *Proceedings of SPIE, 5306*, 416–427.

Prabhakar, S. (2001). *Fingerprint classification and matching using a filterbank*. Ph.D. thesis, Michigan State University, USA.

Qi, M., Lu, Y., Du, N., Wang, C., & Kong, J. (2010). A Novel image hiding approach based on correlation analysis for secure multimodal biometrics. *Journal of Network and Computer Applications, 33*(3), 247–257.

Ratha, N., Connell, J., & Bolle, R. (2001). Enhancing security and privacy in biometric based authentication systems. *IBM Systems Journal, 40*(3), 614–634.

Raval, M., & Rege, P. (2003). Discrete wavelet transform based multiple watermarking scheme. *Proceedings of the Convergent Technologies for the Asia-Pacific Region, 3*, 935–938.

Raval, M., Joshi, M., Rege, P., & Parulkar, S. (2011). Image tampering detection using compressive sensing based watermarking scheme. *Proceedings of MVIP,* 2011.

Rege, P. (2012). *Biometric watermarking*. National Seminar on Computer Vision and Image Processing, Rajkot.

Rohani, M., & Avanaki, A. (2009). A watermarking method based on optimizing SSIM index using PSO in DCT domain. CSICC, pp. 418–423.

Ross, A., & Jain A. (2004). Multimodal biometrics: An overview. *Proceedings of 12th European Signal Processing Conference (EUSIPCO)*, pp. 1221–1224.

Sasidhar, K., Kakulapati, V., Ramakrishna, K., & Ka, K. (2010). Multimodal biometric systems – study to improve accuracy and performance. *International Journal of Computer Science & Engineering Survey (IJCSES), 1*(2), 54–61.

Sheikh, M., & Baraniuk, R. (2007). Blind error free detection of transform domain watermarks. *IEEE International Conference on Image Processing, 5*, V-453.

Shinfeng, D., Shie, S., & Guo, J. (2010). Improving the robustness of DCT based image watermarking against JPEG compression. *Journal of Computer Standards and Interfaces, 32*, 60–67.

Sui, Y., Zou, X., & Du, Y. (2013). Cancellable biometrics. In *Biometrics: From fiction to practice* (pp. 233–252). Singapore: Pan Stanford Publishing Pte. Ltd.

Tagliasacchi, M., Valenzise, G., Tubaro, S., Cancelli, G., & Barni, M. (2009). *A compressive sensing based watermarking scheme for sparse image tampering identification*. Proceedings of ICIP 2009, pp. 1265–1268.

Tamije Selvy, P., Palanisamy, V., & Elakkiya, S. (2013). A novel watermarking images based on wavelet based contourlet transform energized by biometrics. *WSEAS Transactions on Computers, 12*(3), 105–115.

Thanki, R., Kher, R., & Vyas, D. (2011). *Comparative analysis of digital watermarking techniques*. Saarbrücken: LAMBERT Academic Publishing.

Theime, M. (2003). *Multimodal biometric systems: Applications and usage scenarios*. Arlington: Biometric Consortium Conference.

Tiesheng, F., Guiqiang, L., Chunyi, D., & Danhua, W. (2013). A digital image watermarking method based on the theory of compressed sensing. *International Journal Automation and Control Engineering, 2*(2), 56–61.

Tropp, J., & Gilbert, A. (2007). Signal recovery from random measurements via orthogonal matching pursuit. *IEEE Transactions on Information Theory, 53*(12), 4655–4666.

Tsai, H., & Liu, C. (2011). Wavelet based image watermarking with visibility range estimation based on HVS and neural networks. *Pattern Recognition, 44*, 751–763.

Vatsa, M., Singh, R., Mitra, R., & Noore, A. (2004). Digital watermarking based secure multimodal biometric system. *Proceedings of the 2004 I.E. International Conference in Systems Man and Cybernetics, 3*, 2983–2987.

Vatsa, M., Singh, R., & Noore, A. (2005). Improving biometric recognition accuracy and robustness using DWT and SVM watermarking. *IEICE Electronics Express, 1*(12), 362–367.

Vatsa, M., Singh, R., & Noore, A. (2009). Feature based RDWT watermarking for multimodal biometric system. *Image and Vision Computing, 27*(3), 293–304.

Voloshynovskiy, S., Pereira, S., & Pun, T. (2001). Attacks on digital watermarking: Classification, estimation-based attacks and benchmarks. *IEEE Communications Magazine*, 118–126.

Wang, Z., & Bovik, A. (2004). A universal image quality index. *Journal of IEEE Signal Processing Letters, 9*(3), 84–88.

Welling, M. (2005). *Fisher linear discriminant analysis*. Toronto: Department of Computer Science, University of Toronto.

Wolfgang, R., Podilchdc, C., & Dalp, E. (1999). Perceptual watermarks for digital images and video. *Proceedings of IEEE, 87*(7), 1108–1126.

Xu, J., Pang, H., & Zhao, J. (2010). Digital image watermarking algorithm based on fast curvelet transform. *Journal Software Engineering & Applications, 3*, 939–943.

Yang, J., Hua, Y., & William, K. (2000). *An efficient LDA algorithm for face recognition*. Proceedings of the International Conference on Automation, Robotics and Computer Vision (ICARCV 2000), pp. 34–47.

Zebbiche, K., Khelifi, F., & Bouridane, A. (2008). An efficient watermarking technique for the protection of fingerprint images. *EURASIP Journal of Information Security*, 1–20.

Zhang, C., Cheng, L., Zhengding, Q., & Cheng, L. (2008). Multipurpose watermarking based on multiscale curvelet transform. *IEEE Transactions on Information Forensics and Security, 3*(4), 611–619.

Zhang, X., Qian, Z., Ren, Y., & Feng, G. (2011). Watermarking with flexible self-recovery quality based on compressive sensing and compositive reconstruction. *IEEE Transactions on Information Forensics and Security, 6*(4), 1123–1232.

Zhang, Y., Wong, K., Zhang, L., Wen, W., Zhou, J., & He, X. (2015). exploiting random convolution and random subsampling for image encryption and compression. *Signal Processing: Image Communication, 39*(20), 202–211.

Zhang, Y., Zhang, L., Zhou, J., Liu, L., Chen, F., & He, X. (2016). A review of compressive sensing in information security field. *IEEE Access, 4*, 2507–2519.

Zhou, N., Zhang, A., Zheng, F., & Gong, L. (2014). Novel image compression-encryption hybrid algorithm based on key-controlled measurement matrix in compressive sensing. *Optics & Laser Technology, 62*, 152–160.

Zhou, N., Pan, S., Cheng, S., & Zhou, Z. (2016). Image compression – encryption scheme based on hyper-chaotic system and 2D compressive sensing. *Optics & Laser Technology, 82*, 121–133.

Chapter 3
Issues in Biometric System and Proposed Research Methodology

Abstract This chapter presents various issues in existing biometric system, existing multibiometric system, and existing multibiometric watermarking technique. This chapter is given enhancement required in the existing multibiometric system and multibiometric watermarking technique toward research issues. This chapter is also given proposed research methodology.

3.1 Attacks on Biometric or Multibiometric System

Any biometric system is vulnerable to different kinds of attacks which are analyzed by N. Ratha and his research team (Ratha et al. 2001). These attacks are grouped into eight different classes. Figure 3.1 shows the locations of these attacks in a biometric system.

- *Attack Type 1*: Introduced fake or spoof biometric template at sensor level (e.g., fingerprint made from silicon, face mask, a lens with face iris information).
- *Attack Type 2*: Attack on the transmission medium between a sensor and feature extractor module of the system (e.g., the biometric trait is stolen by an impostor).
- *Attack Type 3*: Attack on feature extractor module by using some virus or Trojan horse program (e.g., some features of a biometric trait can be changed by an impostor).
- *Attack Type 4*: Attack on the transmission medium between feature extractor and matcher module (e.g., feature values can be extracted by an impostor).
- *Attack Type 5*: Attack on a matcher module (e.g., the matching score value can be changed by an attacker as per its requirement).
- *Attack Type 6:* Attack on stored templates database (e.g., the biometric template can be tampered or modified by an impostor).
- *Attack Type 7:* Attack on the transmission medium between the stored database and matcher module (e.g., the biometric template can be replaced by an impostor as per its requirement).
- *Attack Type 8:* Attack on decision module (e.g., the decision of identity of a person can be changed by an impostor).

© Springer International Publishing AG 2018

R. M. Thanki et al., *Multibiometric Watermarking with Compressive Sensing Theory*, Signals and Communication Technology,
https://doi.org/10.1007/978-3-319-73183-4_3

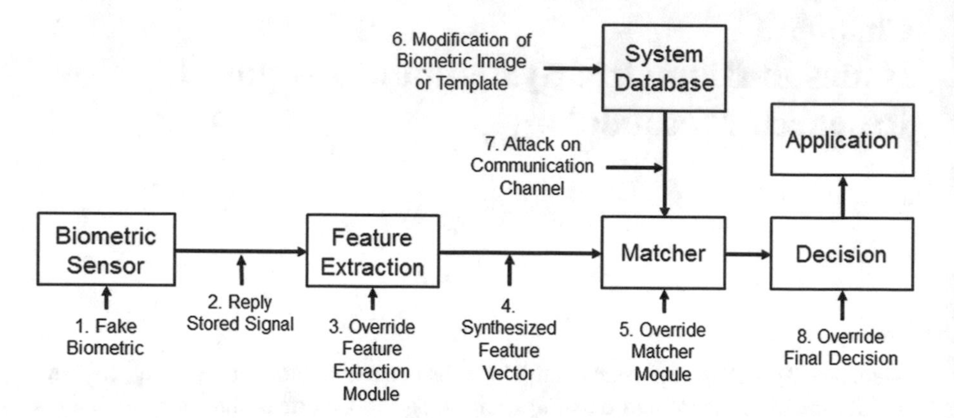

Fig. 3.1 Possible attacks on biometric system

In this book, issues related to security of biometric data or multiple biometric data against *attack type 6* and *attack type 7* are considered. Using the proposed research methodology, protection is provided to biometric data or multiple biometric data against these two attacks.

3.2 Enhancement Required in Existing Multibiometric System and Existing Multibiometric Watermarking Technique

The enhancements required in the existing multibiometric system are mentioned below.

- The protection of biometric image is required at system database in any multibiometric system against spoof or modification attack.
- The protection of biometric image is required at communication channel between two modules of the multibiometric system.

The enhancements required in the multibiometric watermarking technique are mentioned below.

- Few fragile multibiometric watermarking techniques are available for biometric image protection.
- The effect of watermarking on the performance of the multibiometric system is not given in existing multibiometric watermarking techniques.

3.3 Contribution Research Toward Security Enhancement in Multibiometric System and Multibiometric Watermarking Technique

The main contributions of the proposed research toward the society, considering above issues are:

- To propose various watermarking techniques for security of biometric data in multibiometric system.
- The hybrid watermarking technique is designed with a combination of compressive sensing (CS) theory-based encryption procedure which would increase the computational security of watermarking technique.
- To provide the security to biometric data against various modification attacks taken place in the multibiometric system.
- To provide the effect of the proposed watermarking technique on the performance of the multibiometric system.
- The results of the proposed biometric watermarking techniques are compared with the existing watermarking techniques. The advantages and disadvantages of the proposed techniques are also discussed.
- Finally, decision is taken about suitable application of biometric system where these proposed watermarking techniques are suitable.

3.4 Proposed Research Methodology

The watermarking technique is not a new technique for protection of biometric data. In this book, a new watermarking technique with CS theory-based encryption is designed and implemented for multiple biometric data protection. The proposed multibiometric system using watermarking and CS theory-based encryption is shown in Fig. 3.2. This system is used for protection of multiple biometric images in a multibiometric system. This system is analogous to conventional multibiometric system.

In this book, the transform domain-based sparse watermarking technique is used for security of biometric data in multibiometric system. The reason behind choosing this technique is that quantization process and digitalization process are required for generation of encrypted watermark data in a spatial domain-based sparse watermarking technique. The reason for using sparse word in proposed watermarking technique is that encrypted watermark image in terms of sparse data is used instead of its actual pixel information. This proposed system is divided into various procedures such as biometric watermark encryption using CS-based encryption, encrypted biometric watermark embedding procedure, encrypted biometric watermark extracting procedure, biometric watermark decryption using CS- and

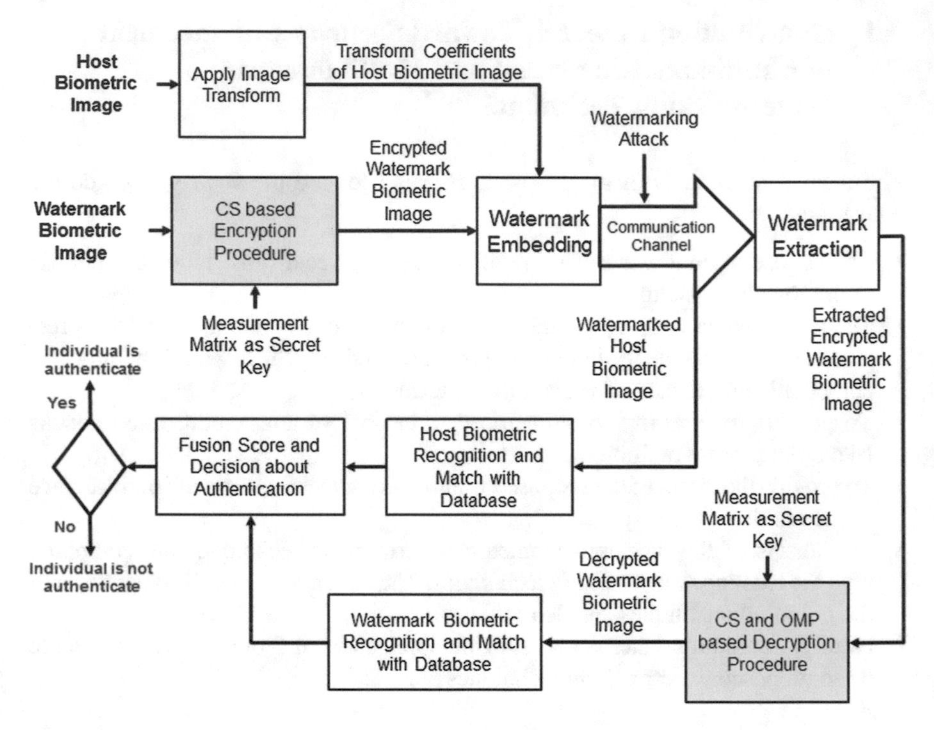

Fig. 3.2 Proposed multibiometric system using watermarking and CS-based encryption

OMP-based decryption, and multiple biometric matching procedures. The details of these procedures are given below.

3.4.1 Watermark Biometric Encryption Using CS-Based Encryption

The steps for watermark biometric encryption using image transform basis matrix and measurement matrix as a secret key are given below.

- Take a watermark biometric image and calculate its size.
- Generated image transform basis matrices using image transform.
- Convert watermark biometric image into its sparse coefficients using image transform basis matrix and its inverted version.

$$x = \Psi_T \times \mathrm{WBI} \times \Psi_T' \tag{3.1}$$

where x is the sparse coefficients of watermark biometric image, Ψ_T is the transform basis matrix, WBI is the original watermark biometric image, and $\Psi_T{}'$ is the inverse transform basis matrix.

- Generated measurement matrix A using Gaussian distribution with mean = 0 and standard deviation = 1. This matrix A is used as secret key for encryption procedure as well as decryption procedure.
- Generate CS measurements of watermark biometric image by multiplying their corresponding sparse coefficients with measurement matrix.

$$y = A \times x \tag{3.2}$$

where y is CS measurements of watermark biometric image in terms of encrypted form and A is a measurement matrix.

- Finally, sampling factor is multiplied with CS measurements of watermark biometric image to generate its sparse data in terms of encrypted form. The sampling factor is user-defined constant value. This factor is not similar to Shannon sampling rate. This factor is used for bringing CS measurements value of watermark biometric image into the range of [−1, 1] for proper watermark embedding. The sampling factor value may be decided by the user as per his/her requirements.

$$W_{\text{Sparse}} = \beta \times y \tag{3.3}$$

where W_{Sparse} is a sparse data of encrypted watermark biometric image, β is a sampling factor, and y is an encrypted watermark biometric image.

- This sparse data of encrypted watermark biometric image are inserted into the host biometric image.

3.4.2 Encrypted Watermark Biometric Embedding Procedure

The steps for encrypted watermark biometric embedding procedure using transform domain watermarking are given below.

- Another biometric image is taken as host biometric image and calculated its size.
- The image transform is applied to host biometric image to get its transform coefficients.
- The sparse data of encrypted watermark biometric image is inserted into transform coefficients of a host biometric image using the watermarking equation (Cox et al. 1997; Shih 2008).

$$\text{WHBI}_{\text{Coefficients}} = \text{HBI}_{\text{Coefficients}} \times \left(1 + k \times W_{\text{Sparse}}\right) \tag{3.4}$$

where $\text{WHBI}_{\text{Coefficients}}$ is modified transform coefficients of host biometric image, $\text{HBI}_{\text{Coefficients}}$ is original transform coefficients of host biometric image, W_{Sparse} is a sparse data of encrypted watermark biometric image, and k is a gain factor.

- The inverse image transform is applied to modified transform coefficients to get a watermarked biometric image at embedder side.

3.4.3 Encrypted Watermark Biometric Extracting Procedure

The steps for encrypted watermark biometric extracting procedure using non-blind transform domain watermarking are given below.

- Take a watermarked biometric image and calculate its size. The image transform is applied to it to get its transform coefficients. Then select transform coefficients which are used for watermark embedding.
- Take an original host biometric image and calculate its size. The image transform is applied to it to get its transform coefficients. Then select transform coefficients which are used for watermark embedding.
- Sparse data of encrypted watermark biometric image is extracted using the reverse procedure of watermark embedding.

$$W_{\text{Extracted}} = \frac{\left(\frac{\text{WHBI}_{\text{Coefficients}}}{\text{HBI}_{\text{Coefficients}}} - 1\right)}{k} \tag{3.5}$$

where $\text{WHBI}_{\text{Coefficients}}$ is transform coefficients of watermarked biometric image, $\text{HBI}_{\text{Coefficients}}$ is original transform coefficients of host biometric image, $W_{\text{Extracted}}$ is extracted sparse data of encrypted watermark biometric image, and k is a gain factor.

- Finally, actual value of encrypted watermark biometric image is taken by extracted sparse data if it is divided by sampling factor value which is used during watermark embedding procedure.

$$y_{\text{Extracted}} = \frac{W_{\text{Extracted}}}{\beta} \tag{3.6}$$

where $y_{\text{Extracted}}$ is an extracted CS measurement of watermark biometric image, $W_{\text{Extracted}}$ is an extracted sparse data of encrypted watermark biometric image, and β is a sampling factor.

3.4.4 *Biometric Watermark Decryption Using CS- and OMP-Based Decryption*

The steps for biometric watermark decryption from extracted encrypted data using CS- and OMP-based decryption are given below.

- Feed the extracted CS measurements of watermark biometric image to OMP algorithm to get its extracted sparse coefficients.

$$x' = \mathrm{OMP}(y_{\mathrm{Extracted}}, A) \tag{3.7}$$

  where x' is extracted sparse coefficients of watermark biometric image, $y_{\mathrm{Extracted}}$ is extracted CS measurements of watermark biometric image, OMP is an orthogonal matching pursuit algorithm, and A is a measurement matrix.
- Obtain actual extracted watermark biometric image from its extracted sparse coefficients using inverse image transform basis matrix and its original version.

$$\mathrm{WBI}' = \Psi_T' \times x' \times \Psi_T \tag{3.8}$$

  where WBI′ is an actual extracted watermark biometric image.
- After getting extracted watermark biometric image, two hypotheses are formulated for authentication of host biometric image.
  - If $\mathrm{SSIM}(\mathrm{IWB}, \mathrm{IWB}_{\mathrm{Reconstructed}}) > \tau$, then host biometric image is authenticated.
  - If $\mathrm{SSIM}(\mathrm{IWB}, \mathrm{IWB}_{\mathrm{Reconstructed}}) < \tau$, then host biometric image is unauthenticated.

  where τ is a predefined threshold value for a decision for authenticity of host biometric images.

3.4.5 *Multiple Biometric Matching Procedures*

In this proposed multibiometric system, watermarked biometric image and extracted watermark biometric image are used for two-level identification of an individual. The steps of two-level identifications are given below.

- An original host biometric image as a query image is compared with a watermarked host biometric image which is stored in the system database. The result is given matching score for watermarked host biometric image.

- An original watermark biometric image as a query image is compared with an extracted watermark biometric image which is stored in the system database. The result is given matching score for extracted watermark biometric image.
- Then an average sum is applied to the matching score of watermarked host biometric image and matching score of extracted watermark biometric image to generate a matching score for the multibiometric system.
- An identity of an individual is possible if a matching score of the multibiometric system is greater than selecting matching score.

3.5 Comparison of Proposed Research Methodology with Conventional Multibiometric Watermarking Technique

In this section, comparison of proposed multibiometric watermarking technique with conventional multibiometric watermarking technique is given. A watermarking technique is embedded in the biometric image as a watermark into another biometric image. This watermarking technique is known as a multibiometric watermarking technique. The block diagram of conventional multibiometric watermarking is shown in Fig. 3.3.

In the conventional multibiometric watermarking technique, watermark biometric image is directly embedded into pixel information or transform coefficients of a host biometric image at the watermark embedder side. On the watermark extraction side, watermark biometric image is extracted from a watermarked biometric image using the reverse procedure of watermark embedding.

In the proposed multibiometric watermarking technique, compressive sensing (CS) theory-based encryption procedure is added in convention watermarking technique. The CS theory-based encryption procedure is added at watermark embedder side. The output of this procedure is an encrypted watermark biometric image. The CS- and OMP-based decryption procedure is added at watermark extraction side. The output of this procedure is decrypted watermark biometric image. The CS

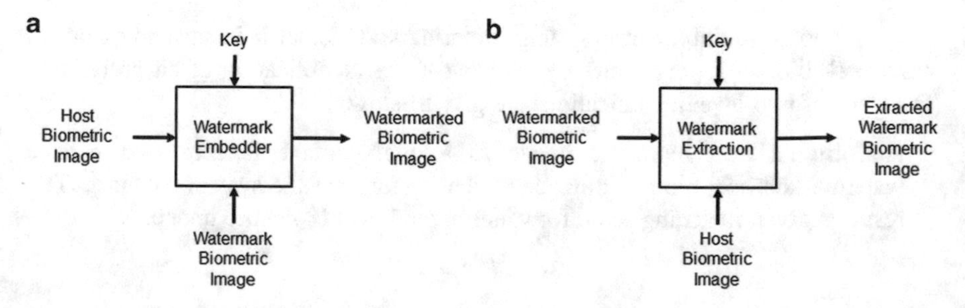

Fig. 3.3 Block diagram of conventional multibiometric watermarking. (**a**) Watermark embedding procedure. (**b**) Watermark extraction procedure

theory-based encryption procedure is providing computational security to watermark biometric image before embedding into host biometric image. The block diagram of proposed multibiometric watermarking is shown in Fig. 3.4.

In watermark embedder side, the watermark biometric image is converted into its sparse coefficients using transform basis matrix and its inverse version. Then CS measurements of a watermark biometric image are generated by multiplication of sparse coefficients and measurement matrix. The CS measurements are represented watermark biometric image into its encrypted form. This procedure of generation of sparse information of watermark biometric image is known as CS theory-based encryption procedure. The sparse data of an encrypted watermark biometric image is generated by multiplying constant value of sampling factor with its CS measurements. This sparse data of an encrypted watermark biometric image is embedded into transform coefficients of a host biometric image. The output of watermark embedded is a watermarked biometric image.

On the watermark extraction side, sparse data of an encrypted watermark biometric image is extracted from a watermarked biometric image using the reverse procedure of watermark embedding. The original transform coefficients of a host biometric image are required for extraction of sparse data of an encrypted biometric image. This is indicated that this proposed multibiometric watermarking technique is non-blind watermarking technique. The sparse data of an extracted encrypted watermark biometric image is divided by sampling factor which is generated at embedder side to get extracted CS measurements of a watermark biometric image. The Orthogonal Matching Pursuit (OMP) (Tropp and Gilbert 2007; Needell 2009) is applied to extracted CS measurements with correct measurement matrix A to get extracted sparse coefficients of a watermark biometric image. These extracted sparse coefficients are multiplied with inverse transform basis matrix and its original version to get extracted watermark biometric image at extraction side.

The comparison of proposed technique with conventional technique shows (based on figures) that proposed technique can provide security to watermark biometric image which is not available into the conventional technique. In this book, various multibiometric watermarking techniques are designed and implemented using various image transforms such as Discrete Wavelet Transform (DWT), Discrete Cosine Transform (DCT), Singular Value Decomposition (SVD), and Fast Discrete Curvelet Transform (FDCT). The proposed watermarking techniques are identified based on transform coefficients of a host biometric image and sparse coefficients of a watermark biometric image used in it. The proposed multibiometric watermarking techniques are given below:

- DWT-based multibiometric watermarking technique: the details of this technique are given in Chap. 4.
- DCT- and DWT-based multibiometric watermarking technique: the details of this technique are given in Chap. 5.
- SVD + DWT- and SVD + DWT-based multibiometric watermarking technique: the details of this technique are given in Chap. 6.

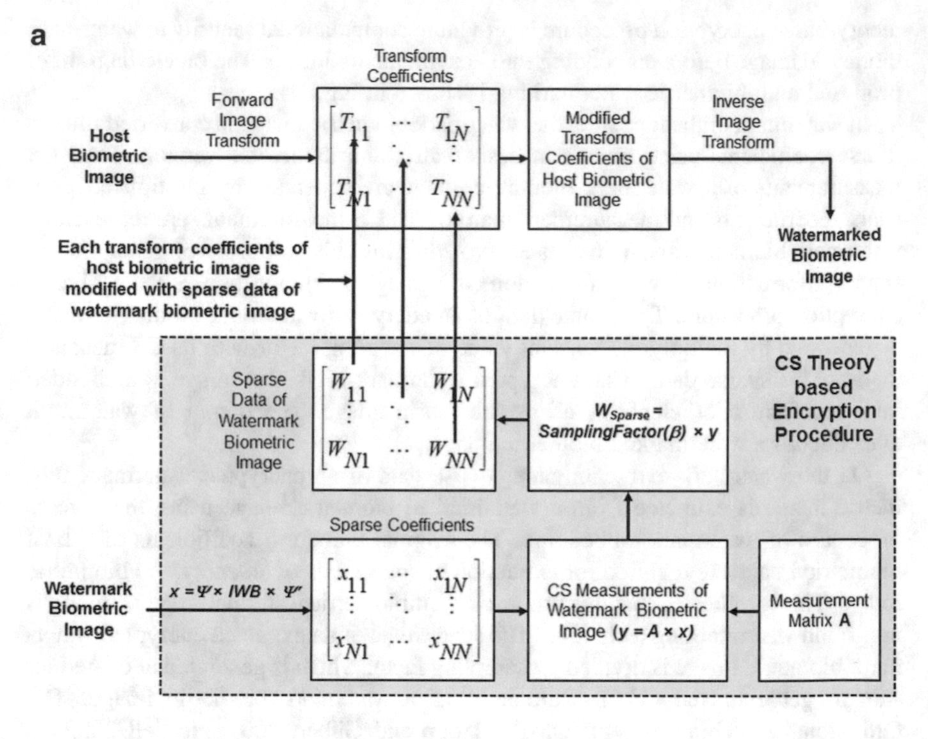

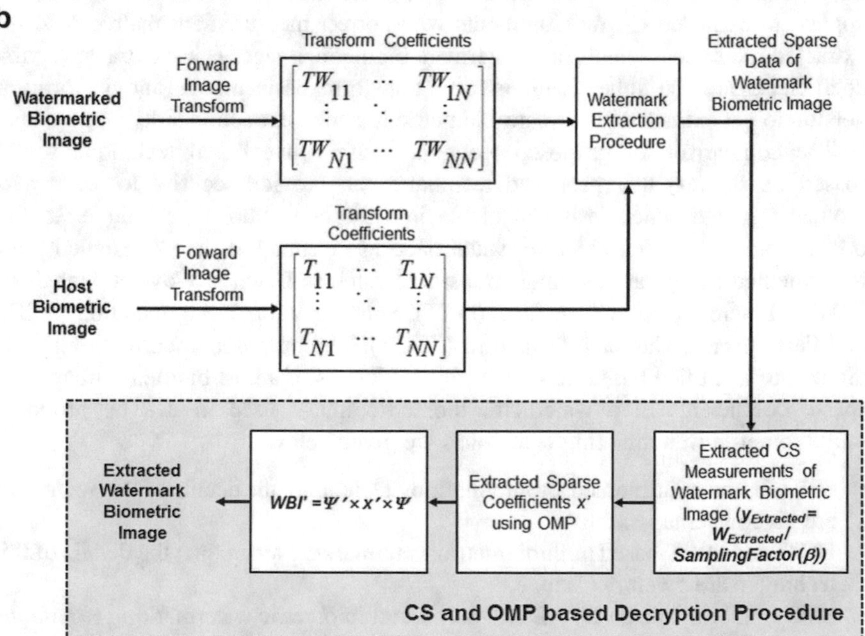

Fig. 3.4 Block diagram of proposed multibiometric watermarking with CS theory-based encryption procedure and decryption procedure. (**a**) Proposed watermarking embedding procedure. (**b**) Proposed watermarking extracting procedure

- FDCuT- and DCT-based multibiometric watermarking technique: the details of this technique are given in Chap. 7.

3.6 Resources Used for Implementation of Proposed Multibiometric Watermarking Technique

In this book, the face image of an individual is used as a host biometric image, and fingerprint image of an individual is used as a watermark biometric image. The reason behind choosing these two biometric characteristics is that their are easily available and unique for every person. The size of these two images is 128 × 128 pixels and .bmp format with different grayscale. In the MATLAB 2013a platform, image processing toolbox and wavelet toolbox is used for implementation of proposed multibiometric watermarking techniques. The implementation of the all proposed techniques is done on the laptop with 2 GHz Core 2 Duo processor and 2 GB RAM.

The analysis of the effect of all proposed techniques on the multibiometric system is done using individual watermarked face-based system, individual extracted watermark fingerprint-based system, and face-fingerprint-based multibiometric system. For analysis purposes of proposed techniques, standard fingerprint images are taken from Fingerprint Verification Competition (FVC) database 2004. The standard face images are taken from an Indian face database (Jain and Mukherjee 2002), FEI face database, and CVL face database. More information about face database and fingerprint database are given below.

3.6.1 Fingerprint Verification Competition (FVC) Fingerprint Database (FVC Fingerprint Database, 2004)

This fingerprint database has four different databases. The databases are named as DB1, DB2, DB3, and DB4. These databases are collected using different sensors and technologies. This fingerprint database contains the fingerprint of students whose average age is around 24 years. These students are enrolled in the computer science degree program at the Bologna University, Italy. Some of the characteristics of a fingerprint presented in this database are given below:

- Students are 3 groups of 30 persons. Each group was associated with a DB. Every fingerprint is acquired using different fingerprint scanners.
- No techniques or efforts are made to control image quality. The sensor plates are not cleaned systematically.

Table 3.1 Information of FVC fingerprint database (FVC 2002a and FVC 2004)

Database	Sensor type	Image size	Set A (images)	Set B (images)	Resolution
DB1	Optical sensor	388 × 374 (142 K pixels)	800	80	500 dpi
DB2	Optical sensor	296 × 560 (162 K pixels)	800	80	569 dpi
DB3	Capacitive sensor	300 × 300 (88 K pixels)	800	80	500 dpi
DB4	Synthetic fingerprint generator software	288 × 384 (108 K pixels)	800	80	About 500 dpi

This database is divided into two sets such as set *A* and set *B*. All the information on this fingerprint database is given in below Table 3.1.

In this book, 80 images are taken from the FVC2004 DB4 setB fingerprint database, and 80 images are taken from FVC2002 DB3 setB fingerprint database. These images are used as authentic fingerprint images. Also, 80 images are taken from FVC2002 DB4 setB, and 80 images are taken from FVC2004 DB3 setB. These images are used as fake fingerprint images. Figure 3.5 shows few samples fingerprint images which are used in this book.

3.6.2 Indian Face Database (Jain and Mukherjee 2002)

This face database is taken by V. Jain and A. Mukherjee in February 2002 at the IIT Kanpur campus. This database contains a set of different face images. In this database, 11 different images of 61 persons are available. The database contains 671 face images with different orientations. The images are colored and in JPEG format. The size of each image is 640 × 480 pixels.

The face images are divided into two main directories like females and males. In female's directories, there are images of 22 different females with the name as a serial number is available. In male's directories, there are images of 39 different males with the name as the serial number is available. There are 11 different images of that individual, which have names of from mvc-ABCf.jpg, where ABC is the image number of that individual. These 11 images are having different orientations of the face like looking front, looking left, looking right, looking up, looking up toward the left, looking up toward the right, and looking down. Some of the emotions are available in images like neutral, smile, laughter, and sad.

In the book, 50 images are taken from this database. These images are used as an authentic face image. For test of proposed techniques, images are converted into grayscale with .bmp format, and Fig. 3.6 shows some sample images from the Indian face database (in grayscale).

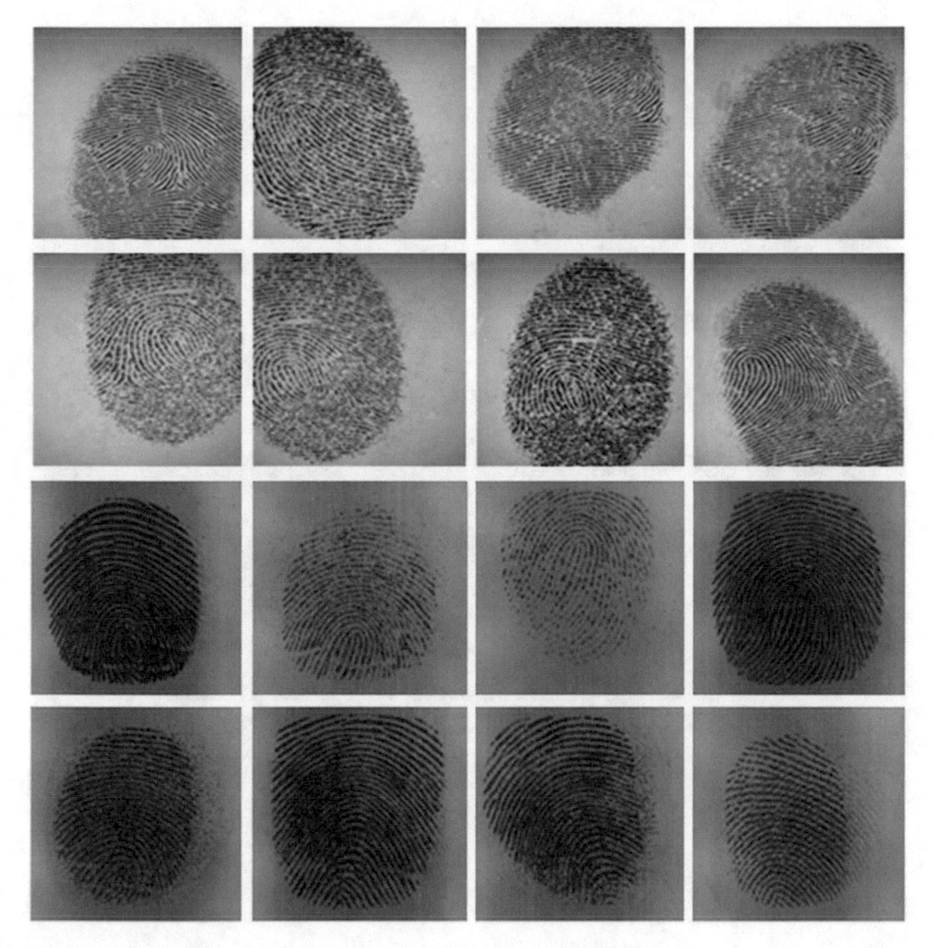

Fig. 3.5 Samples of fingerprint images from FVC fingerprint database

3.6.3 FEI Face Database (FEI Face Database 2005, 2006)

This database is a Brazilian face database. This database contains a set of face images taken by the Artificial Intelligence Laboratory of FEI in Sao Bernardo do Campo, Sao Paulo, Brazil, during June 2005 to March 2006. There are 14 images for 200 persons available in the database. The database contains 2800 face images with different orientations.

All images are colored and taken against a white background. The size of each face image is 640 $\times$ 480 pixels. These face images are an upright frontal position with rotation up to about 180 degrees. The database contains images from FEI students and staff between the ages of 19 and 40 years old with different hairstyles and appearance. The database contains 100 male and 100 female individual face images.

Fig. 3.6 Sample face images from the Indian face database

In this book, 160 images are taken from this database, where 110 images are used as authentic face images and 50 images are used as fake face images. For test of proposed techniques, images are converted into grayscale with .bmp format, and Fig. 3.7 shows some sample images from FEI face database (in grayscale).

3.6.4 CVL Face Database (CVL Face Database 2003; Solina et al. 2003)

This database is a Slovenia face database. This database contains a set of face images taken by a faculty of computer and information science department, Ljubljana University, Slovenia. There are 7 images for 114 persons available in the database.

Fig. 3.7 Sample face images from FEI face database

All images are color and taken against a white background. The size of each face image is 640 × 480 pixels.

The database contains images from the students with age of around 18 years old with different positions such as far left side view, side view with 45 angles, side view with 135 angles, far right side view, front view with a serious expression, frontal view with a smile showing no teeth, and frontal view with a smile showing teeth. These face images are taken using the Sony digital camera under uniform illumination and no flash. The database contains 90% images of male individuals.

In this thesis, 110 images are taken from this database. These images are used as fake face images. For test of proposed techniques, images are converted into grayscale with .bmp format, and Fig. 3.8 shows some sample images from CVL face database (in grayscale).

Fig. 3.8 Sample face images from CVL face database

References

Cox, I., Kilian, J., Shamoon, T., & Leighton, F. (1997). Secure spread spectrum watermarking for multimedia. *IEEE Transactions on Image Processing, 6*(12), 1673–1687.

FEI Face Database. (2005, 2006). Available http://fei.edu.br/~cet/facedatabase.html

FVC Fingerprint Database. (2002, 2004). Available http://csr.unibo.it/fvc2004/, http://csr.unibo.it/fvc2002/

Jain, V., & Mukherjee, A. (2002). Indian Face Database. (2002). Available http://vis-www.cs.umass.edu/~vidit/IndianFaceDatabase

Needell, D. (2009). *Topics in compressed sensing*. Ph.D. thesis, University of California, USA.

Peer, P. (2003). *CVL Face Database*. Available http://www.lrv.fri.uni-lj.si/facedb.html

Ratha, N., Connell, J., & Bolle, R. (2001). Enhancing security and privacy in biometric based authentication systems. *IBM Systems Journal, 40*(3), 614–634.

Shih, F. (2008). *Digital watermarking and steganography – fundamentals and techniques* (pp. 39–41). Boca Raton: CRC Press.

Solina, F., Peer, P., Batagelj, B., Juvan, S., & Kova, J. (2003). Color-based face detection in the '15 seconds of fame' art installation. In: *Mirage 2003, Conference on Computer Vision/Computer Graphics Collaboration for Model-based Imaging, Rendering, Image Analysis and Graphical special Effects, March*, pp. 38–47.

Tropp, J., & Gilbert, A. (2007). Signal recovery from random measurements via orthogonal matching pursuit. *IEEE Transactions on Information Theory, 53*(12), 4655–4666.

Chapter 4
Multibiometric Watermarking Technique Using Discrete Wavelet Transform (DWT)

Abstract This chapter presents technical details and sparsity property of Discrete Wavelet Transform (DWT). The multibiometric watermarking technique using DWT is explained and analyzed in this chapter. The comparison of presented watermarking technique with existing watermarking techniques is also given in this chapter.

4.1 Discrete Wavelet Transform (DWT)

The Discrete Wavelet Transform (DWT) is used for converting an image into its frequency coefficients. It is used for multi-resolution analysis where different frequency coefficients are analyzed. The advantage of a wavelet transform is to decompose an image into various frequency subband coefficients. These subband coefficients are known as LL subband, LH subband, HL subband, and HH subband.

Equations 4.1 and 4.2 show DWT which converted an image into an approximation and detail coefficients of its wavelet. Equation 4.3 shows the reverse procedure for image reconstruction from its wavelet coefficients (Lopez and Boulgouris 2010).

$$W_{\varphi}(j_0, k) = \frac{1}{\sqrt{M}} \sum\nolimits_{n} f(n) \cdot \phi_{j_0,k}(n) \tag{4.1}$$

$$W_{\Psi}(j, k) = \frac{1}{\sqrt{M}} \sum\nolimits_{n} f(n) \cdot \Psi_{j,k}(n) \tag{4.2}$$

For $j \geq j_0$,

$$f(n) = \frac{1}{\sqrt{M}} \sum\nolimits_{n} f(n) \cdot \phi_{j_0,k}(n) + \frac{1}{\sqrt{M}} \sum\nolimits_{n} f(n) \cdot \Psi_{j,k}(n) \tag{4.3}$$

The DWT is calculated by successive low-pass and high-pass filtering of the image in different frequency coefficients. At each wavelet-level decomposition, the high-pass filter produces detail coefficient and the low-pass filter produces

© Springer International Publishing AG 2018

R. M. Thanki et al., *Multibiometric Watermarking with Compressive Sensing Theory*, Signals and Communication Technology,
https://doi.org/10.1007/978-3-319-73183-4_4

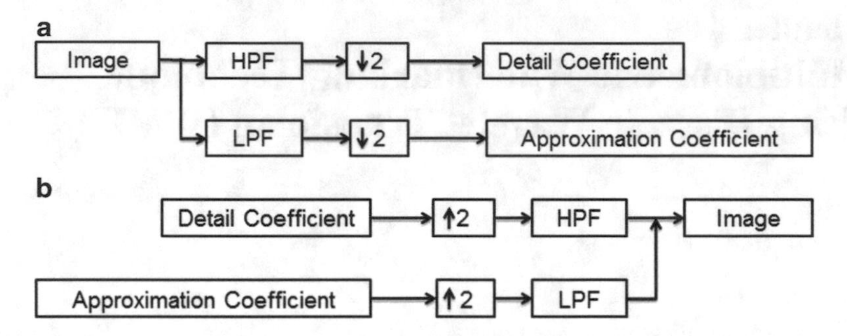

Fig. 4.1 Analysis and synthesis process of Discrete Wavelet Transform (DWT). (**a**) Analysis. (**b**) Synthesis

approximation coefficient with association to scaling function. The analysis and synthesis procedure for DWT is shown in Fig. 4.1 (Lopez and Boulgouris 2010).

The level of synthesis and analysis procedures must be the same for successive reconstruction of the image. To achieve perfect reconstruction of the image, the analysis and synthesis filters have to satisfy certain conditions (Lopez and Boulgouris 2010). If filters satisfy these conditions, then the image is reconstructed perfectly.

$$G_0(z) \cdot H_0(z) + G_1(z) \cdot H_1(z) = 2 \tag{4.4}$$

$$G_0(z) \cdot H_0(-z) + G_1(z) \cdot H_1(-z) = 0 \tag{4.5}$$

where, H_0 and H_1 represent low-pass and high-pass analysis filters and G_0 and G_1 represent low-pass and high-pass synthesis filters.

The wavelet transform of an image is decomposed equal to the actual size of the image, but the representation of coefficients is different compared to Discrete Cosine Transform (DCT). Every wavelet level produces different coefficients which come from a set of different combinations of high- and low-pass filters. The approximation coefficient of the image is the result of low-pass filtering. The detail coefficient of the image contains vertical, horizontal, or diagonal detail on the filters applied in each direction. The subsequent level of the wavelet decomposition of an image is obtained after applying onto the approximation coefficient of an image from the previous level (Lopez and Boulgouris 2010). This procedure generated new detail and approximation coefficients of an image. Figure 4.2 shows a different level of the wavelet decomposition of the image.

The important properties of DWT are given below. The DWT can be used for generation of sparse measurements of any image (Gonzales and Woods 2002; Jain 1999). The examples of DWT are Haar, Symlet, Daubechies, and Biorsplines.

1. The DWT is real and orthogonal.
2. The DWT is fast transform.

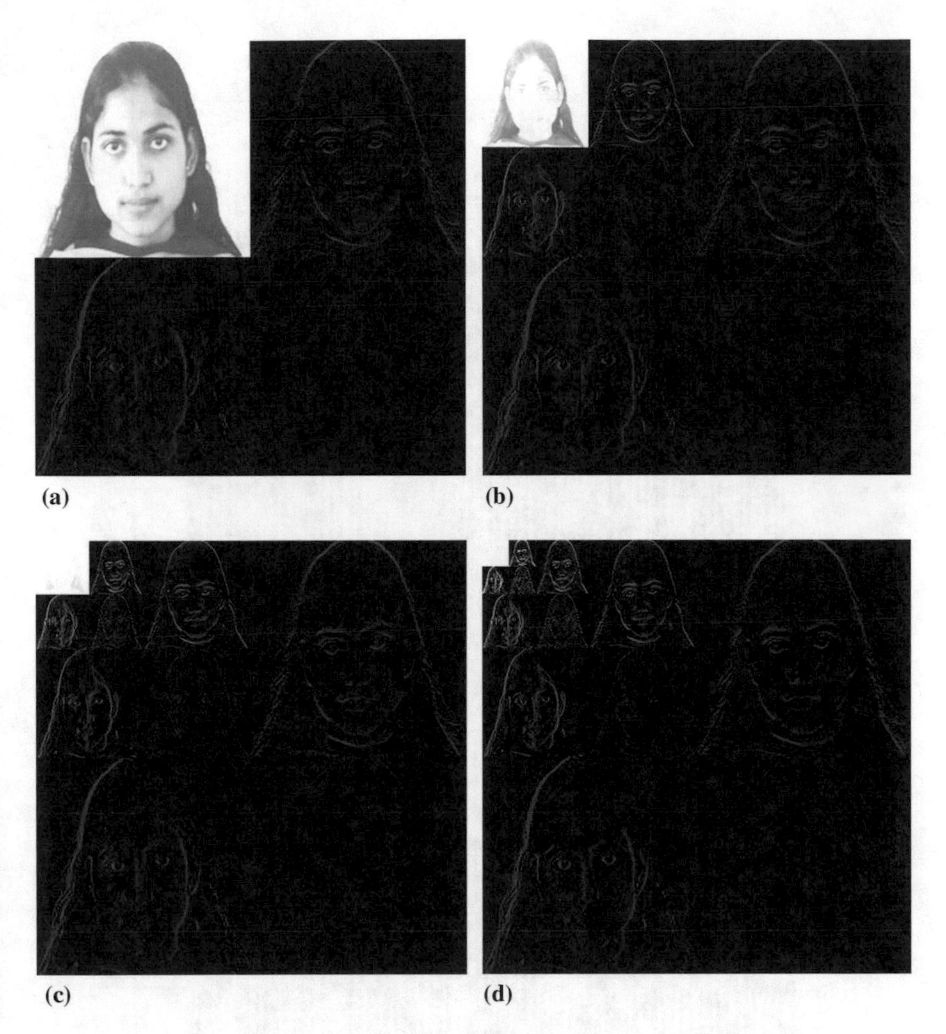

Fig. 4.2 Different level decompositions of DWT. (**a**) First level DWT decomposition. (**b**) Second level DWT decomposition. (**c**) Third level DWT decomposition. (**d**) Fourth level DWT decomposition

3. When DWT is applied to an image, then the image is converted into approximation and detail wavelet coefficients. The detail wavelet coefficients and approximation wavelet coefficients are shown in Fig. 4.3.

The sparse coefficients of a watermark biometric image are generated using first level 2D Discrete Wavelet Transform (DWT). The detail wavelet coefficients of a watermark biometric image are chosen as sparse coefficients in this chapter. The reason behind choosing detail wavelet coefficients as sparse coefficients is that these coefficients are sparser than approximation wavelet coefficients.

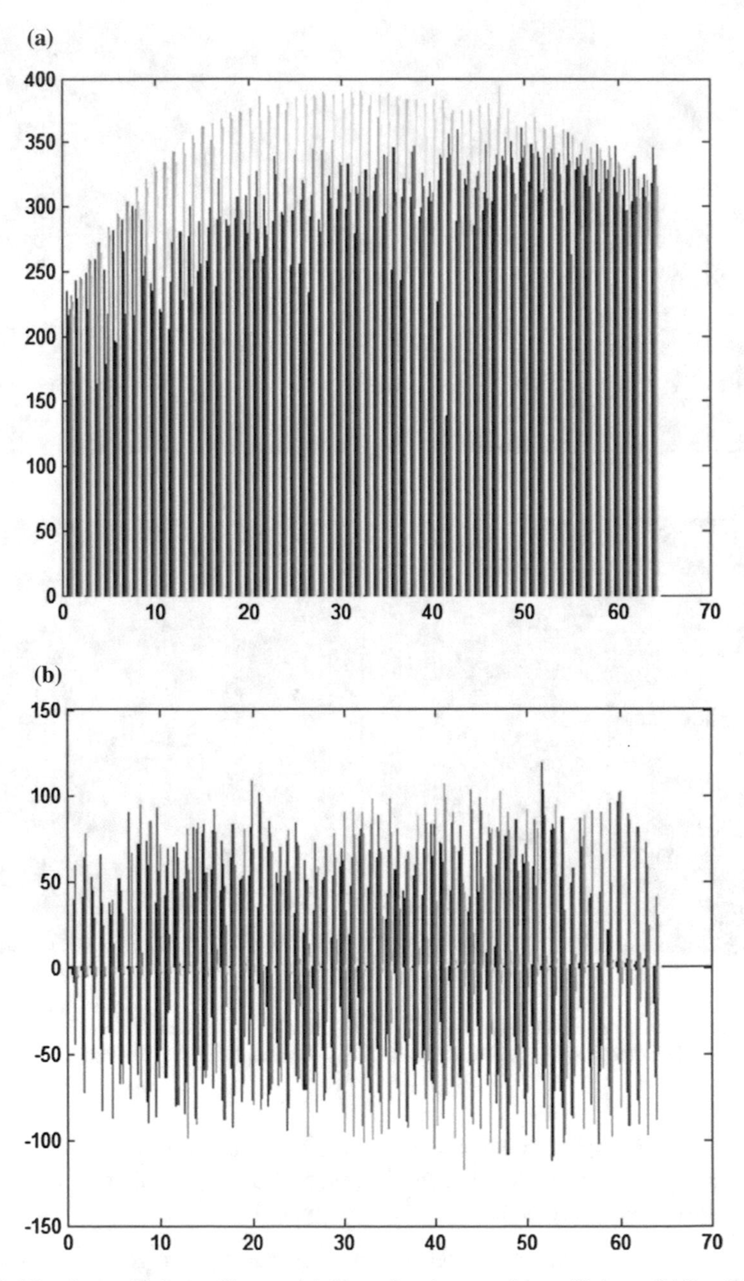

Fig. 4.3 Wavelet coefficients of image. (**a**) Approximation wavelet coefficients. (**b**) Detail wavelet coefficients

In this book, a wavelet basis matrix-based method is used for generation of sparse coefficients of a watermark biometric image. In this method, all wavelet coefficients of watermark biometric image are considered as sparse coefficients. This method is used in proposed multibiometric watermarking technique in Chaps. 5 and 6. In this method, the wavelet basis matrix generated using the wavelet matrix method is described by Vidakovic (1999) and Jie Yan (2009). The steps for wavelet basis matrix generation in MATLAB are given below:

- Generate sparse wavelet coefficients H and G using any wavelet filters.

$$[H, G] = w\text{filters}\left('\text{wavename}', d\right); \tag{4.6}$$

where wave name is wavelet filter name such as db1, db2, sym2, sym4, and sym8.

- Choose the size of wavelet matrix to be generated. This wavelet matrix N should be a power of 2.
- Then generate a circular matrix with a size of $N/2$ for low-pass wavelet coefficients and high-pass wavelet coefficients using the below procedure.

$$\begin{aligned}
&L = \text{length}(H);\\
&\text{Max} = \text{log2}(N);\\
&\text{Min} = \text{log2}(L) + 1;\\
&\text{for}, j = \text{Min: Max}\\
&N = 2^j;\\
&\text{for}, i = 1 : N/2\\
&H1(i,:) = \text{circshift}(H', 2*(i-1))';\\
&G1(i,:) = \text{circshift}(G', 2*(i-1))';\\
&\text{end}
\end{aligned} \tag{4.7}$$

- Then wavelet coefficients with the size of N are obtained by combining low-pass and high-pass wavelet coefficients.

$$W = [H1; G1]; \tag{4.8}$$

- Finally, these wavelet coefficients are multiplied by the identity matrix with a size of N to get one row of the wavelet matrix with a size of N.
- Then perform this procedure N times to get a wavelet basis matrix with a size of $N \times N$ where row data in wavelet matrix are shifting two positions on the right side in every iteration.
- The output of this procedure in MATLAB is given in Fig. 4.4.

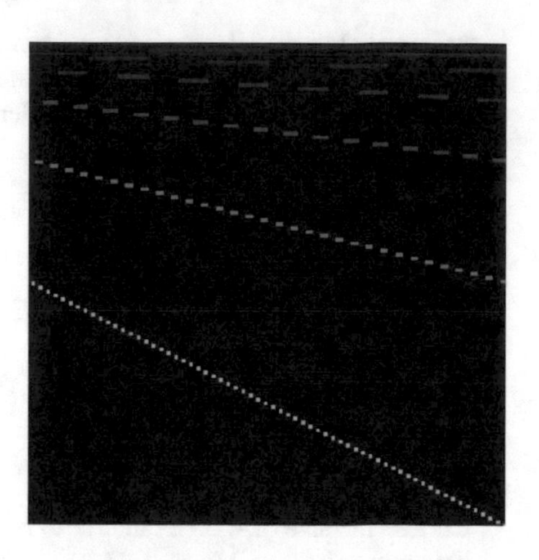

Fig. 4.4 DWT basis matrix

4.2 Multibiometric Watermarking Technique Using DWT

In this proposed watermarking technique, watermarking is performed in DWT domain, while encryption of watermark biometric image is performed using DWT and CS theory-based encryption process. The DWT is used for generation of sparse coefficients of the watermark biometric image. The sparse data of an encrypted watermark biometric image is inserted into approximation wavelet coefficients of a host biometric image. This proposed technique is non-blind technique because the original host biometric image is required at extraction side.

In the presented multibiometric watermarking technique, fourth level approximation coefficients of a host biometric image are modified according to gain factor and values of sparse data of encrypted watermark biometric image. The reason behind choosing fourth level wavelet coefficients is that it has a very small amount of low-frequency coefficients and is not used for watermarking by researchers. Also, the size of these coefficients is equal to the size of sparse measurements of the watermark biometric image. This proposed multibiometric watermarking technique is divided into two procedures such as watermark biometric encryption and embedding of encrypted watermark biometric and extraction of encrypted watermark biometric and decryption of watermark biometric. The proposed block diagram of multibiometric watermarking technique using DWT is shown in Fig. 4.5.

4.2.1 Watermark Biometric Image Encryption Procedure and Embedding Procedure of Encrypted Watermark Biometric Image

The steps for watermark biometric image encryption procedure and embedding procedure of encrypted watermark biometric image are given below:

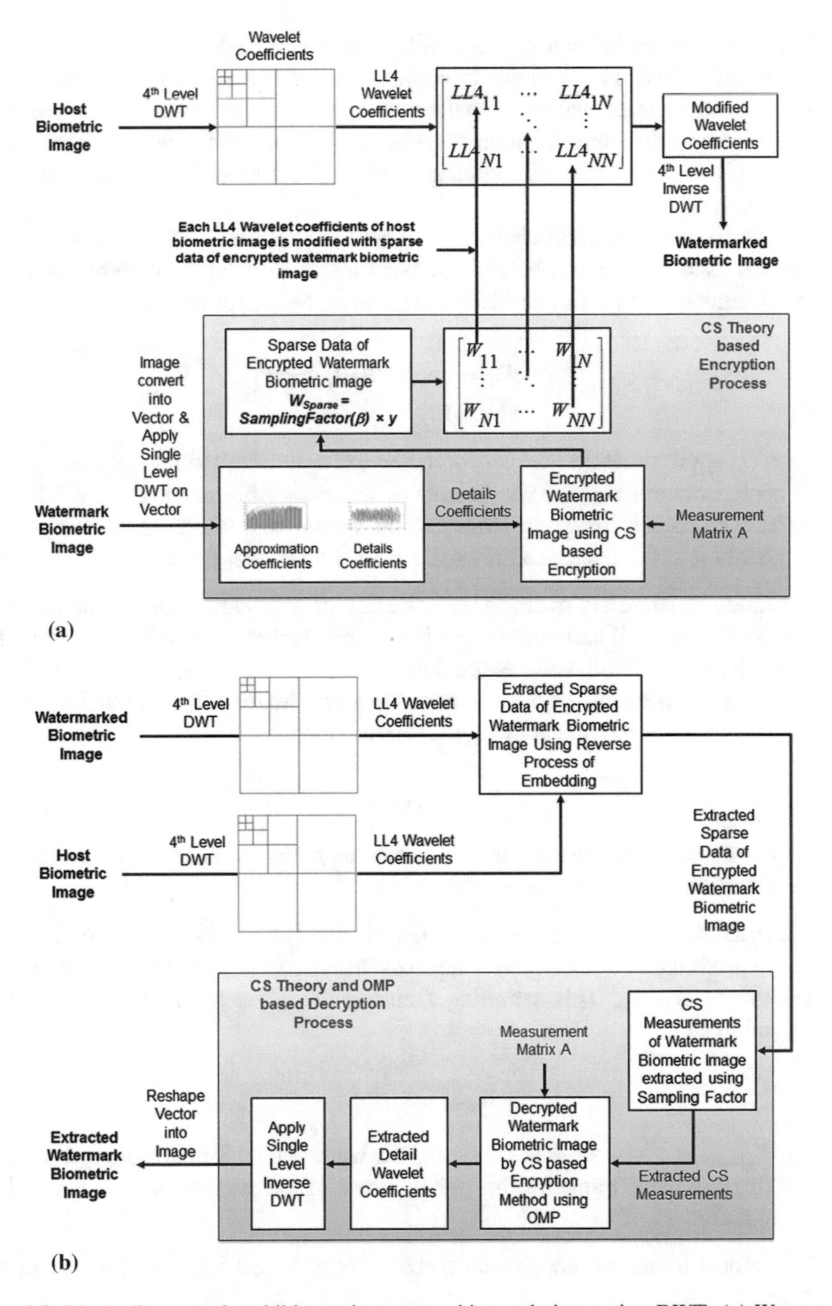

Fig. 4.5 Block diagram of multibiometric watermarking technique using DWT. (**a**) Watermark biometric encryption and embedding of encrypted watermark biometric. (**b**) Extraction of encrypted watermark biometric and decryption of watermark biometric

- Take a watermark biometric image WBI with a size of $N \times N$ and calculate the size of the image. The watermark biometric image is converted into vector.
- The first level Haar wavelet transform is applied on the vector. The vector is decomposed into approximation wavelet coefficients and detail wavelet coefficients of a watermark biometric image after application of the first level wavelet transform.
- The detail wavelet coefficients are chosen as sparse coefficients which are denoted as x. The reason behind choosing detail wavelet coefficients as sparse coefficients is that detail coefficients are sparser than approximation coefficients.

$$\begin{aligned} [\mathrm{cA}, \mathrm{cD}] &= \mathrm{DWT}(\mathrm{WBI}, '\mathrm{haar}') \\ \mathrm{cD} &= x \end{aligned} \tag{4.9}$$

where cA is approximation wavelet coefficients of watermark biometric image, cD is detail approximation wavelet coefficients of watermark biometric image, WBI is a watermark biometric image in terms of the vector, x is a sparse coefficients of watermark biometric image, and Haar is a name of wavelet filter.

- Generate measurement matrix A with the size of $N \times N$ using a normal distribution with mean = 0 and variance = 1. This measurement matrix is the same for embedder side as well as for extraction side.
- The CS measurements of a watermark biometric image are generated by multiplying its sparse coefficients with the measurement matrix.

$$y = A \times x \tag{4.10}$$

where y is an encrypted watermark biometric image in terms of CS measurements and A is a measurement matrix.

- The encrypted watermark biometric image is multiplied with a sampling factor to get a sparse data of encrypted watermark biometric image. This sparse data is denoted as W_{Sparse}. This sampling factor is the same for embedder side and extraction side.

$$W_{\mathrm{Sparse}} = \beta \times y \tag{4.11}$$

where W_{Sparse} is a sparse data of encrypted watermark biometric image, y is an encrypted watermark biometric imager in terms of CS measurements, and β is a sampling factor.

- Take a host biometric image with a size of $N \times N$ and calculate the size of the image.
- The fourth level Haar wavelet transform is applied to host biometric image to get various wavelet subbands such as LL4, HL4, LH4, and HH4.

- The sparse data of an encrypted watermark image is inserted into the fourth level LL subband coefficients of a host biometric image using multiplicative watermarking equation (Cox et al. 1997; Shih 2008).

$$\mathrm{WLL4} = \mathrm{LL4}^{*}\left(1 + k \times W_{\mathrm{Sparse}}\right) \quad (4.12)$$

where WLL4 is the modified fourth level LL subband coefficients of host biometric image, LL4 is the original fourth level LL subband coefficients of host biometric image, W_{Sparse} is sparse data of encrypted watermark biometric image, and k is a gain factor.

- The fourth level inverse Haar wavelet transform is applied to modified wavelet coefficients with original wavelet coefficients of host biometric image to get the watermarked biometric image.

4.2.2 *Extraction Procedure of Encrypted Watermark Biometric Image and Decryption Procedure for Watermark Biometric Image*

The steps for extraction procedure of encrypted watermark biometric image and decryption procedure for watermark biometric image are given below:

- Take a watermarked biometric image which may be corrupted or degraded by the impostor. The fourth level Haar wavelet transform is applied to watermarked biometric image to get wavelet coefficients of various subbands such as LLW4, HLW4, LHW4, and HHW4.
- Take an original host biometric image. The fourth level Haar wavelet transform is applied to host biometric image to get wavelet coefficients of various subbands such as LL4, HL4, LH4, and HH4.
- The sparse data of encrypted watermark biometric image is extracted using the reverse procedure of embedding.

$$W_{\mathrm{Extracted}} = \frac{\left(\frac{\mathrm{WLL4}}{\mathrm{LL4}} - 1\right)}{k} \quad (4.13)$$

where WLL4 is the fourth level LL subband coefficients of watermarked biometric image, LL4 is the original fourth level LL subband coefficients of host biometric image, $W_{\mathrm{Extracted}}$ is extracted sparse data of encrypted watermark biometric image, and k is a gain factor.

- The extracted sparse data of encrypted watermark biometric image is divided by a sampling factor to get extracted encrypted watermark biometric image.

$$y_{\text{Extracted}} = \frac{W_{\text{Extracted}}}{\beta} \tag{4.14}$$

where $W_{\text{Extracted}}$ is extracted sparse data of encrypted watermark biometric image, $y_{\text{Extracted}}$ is extracted encrypted watermark biometric image, and β is a sampling factor.

- After extracting encrypted watermark biometric image, the decryption of watermark biometric image process is performed using CS theory and Orthogonal Matching Pursuit (OMP) algorithm.
- The OMP is applied to extracted encrypted watermark biometric image with correct measurement matrix to get sparse coefficients. These coefficients are detail wavelet coefficients of a watermark biometric image.

$$x' = \text{OMP}(y_{\text{Extracted}}, A) \tag{4.15}$$

where x' is an extracted detail wavelet coefficient of watermark biometric image.

- The first level inverse Haar wavelet transform is applied to extracted detail wavelet coefficients with the original approximation wavelet coefficients to get the actual pixel values of watermark biometric image in terms of the vector.

$$\text{RI} = \text{IDWT}\left(cA, x', '\text{haar}'\right) \tag{4.16}$$

where RI is an extracted pixel value of watermark biometric image in terms of the vector, IDWT is an inverse DWT, and Haar is a name of wavelet filter.

- Finally, reshape the vector into the matrix to get extracted watermark biometric image at extraction side.

$$\text{EWBI} = \text{reshape}(R, M, N) \tag{4.17}$$

where EWBI is an extracted watermark biometric image, M is a row size of watermark biometric image, and N is column size of watermark biometric image.

- After extracting watermark biometric image, two hypotheses are formulated for authentication of host biometric image.
  - If SSIM(WBI, EWBI) $> \tau$, then host biometric image is authenticated.
  - If SSIM(WBI, EWBI) $< \tau$, then host biometric image is unauthenticated.

where τ is a predefined threshold value for a decision about an authenticity of host biometric image.

4.3 Experimental Results

For testing and analysis of this proposed technique, 8-bit-grayscale 50 face images from Indian face database and 110 face images from FEI face database are taken as host biometric images. The 8-bit-grayscale 80 fingerprint images from FVC2002 DB3 setB and 80 fingerprint images from FVC2004 DB4 setB are taken as watermark biometric images. The size of host face image and watermark fingerprint image is 128×128 pixels. The sparse data of encrypted watermark fingerprint image is inserted into host face image using the below procedure.

First, watermark fingerprint image is converted into the vector. The first level Haar wavelet transform is applied to the vector to get detail wavelet coefficients and approximation wavelet coefficients. The detail wavelet coefficients of watermark fingerprint image are taken as sparse coefficients x with a size of 8192×1. The measurement matrix A with the size of 64×8192 is generated using a normal distribution with mean $= 0$ and variance $= 1$. The CS measurements of watermark fingerprint image with a size of 64×1 are generated using $y_{64\times1} = A_{64\times8192} \times x_{8192\times1}$. The encrypted watermark fingerprint image is multiplied with sampling factor to get sparse data of encrypted watermark fingerprint image. Finally, this sparse data is converted into the 8×8 size of the matrix and denoted as W_{Sparse}.

The sparse data W_{Sparse} of encrypted watermark fingerprint image is inserted into the fourth level LL wavelet coefficients of host face image using gain factor at embedder side. On the extraction side, sparse data $W_{\text{Extracted}}$ of encrypted watermark fingerprint image is extracted using reverse procedure of embedding. The decryption of watermark fingerprint image from extracted encrypted fingerprint image can be done by using the below procedure at extraction side.

For decryption of watermark fingerprint image, the input of OMP algorithm is encrypted watermark fingerprint image with a size of 64×1 and measurement matrix A with a size of 64×8192. The output of OMP algorithm is sparse coefficients of watermark fingerprint image with a size of 8192×1. The inverse first level Haar wavelet transform is applied to these extracted sparse coefficients with original approximation wavelet coefficients to get extracted watermark fingerprint image in terms of the vector. Finally, reshape this vector into matrix to get extracted watermark fingerprint image at extraction side.

Figure 4.6 shows the original face image and watermarked face image, original fingerprint image and extracted fingerprint image, and encrypted fingerprint image and extracted fingerprint image. These results are generated using gain factor k value 0.2 and sampling factor β value 0.01. For testing of nature of proposed watermarking technique, various watermarking attacks such as JPEG compression with different quality factor ($Q = 90$–50), the addition of noise (Gaussian noise, speckle noise, salt and pepper noise), applied different image filter with different filter mask size

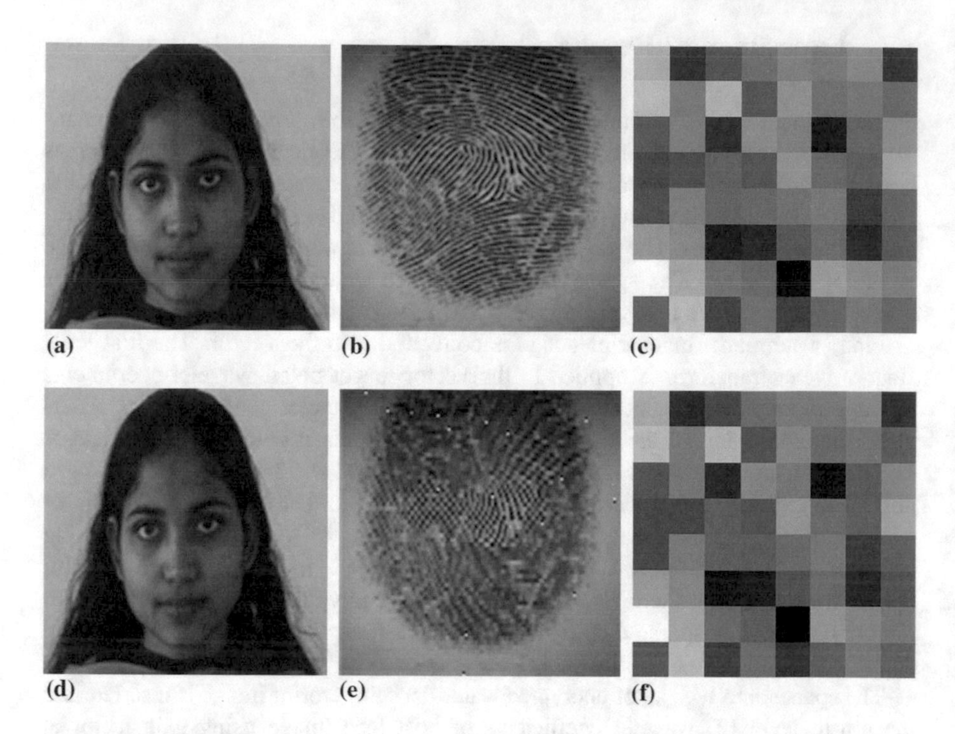

Fig. 4.6 Experimental results of multibiometric watermarking technique using DWT for gain factor $k = 0.2$. (**a**) Host face image. (**b**) Watermark fingerprint image. (**c**) Encrypted watermark fingerprint image. (**d**) Watermarked face image. (**e**) Extracted watermark fingerprint image. (**f**) Extracted encrypted watermark fingerprint image

(median filter, mean filter, and Gaussian low-pass filter), sharpening, blurring, histogram equalization, and different geometric attacks like flipping, rotation, and cropping are applied on watermarked face image. The performance result of watermarking attacks on proposed watermarking technique is shown in Fig. 4.7.

The quality measure such as peak signal to noise ratio (PSNR) is used for quality check between original face and watermarked face image. The structural similarity index measure (SSIM) is used for quality check between the original watermark fingerprint and extracted watermark fingerprint image in proposed multibiometric watermarking technique. The results of multibiometric watermarking technique using DWT under various watermarking attacks are summarized in Table 4.1.

For authentication of the host face image of an individual, SSIM value between the watermark fingerprint image and extracted watermark fingerprint image must be greater than the predefined threshold value $\tau = 0.90$. SSIM values in Table 4.1 are indicated that when a watermarking attack is applied on watermarked face image, then the watermark fingerprint image is extracted successfully at extraction side, and

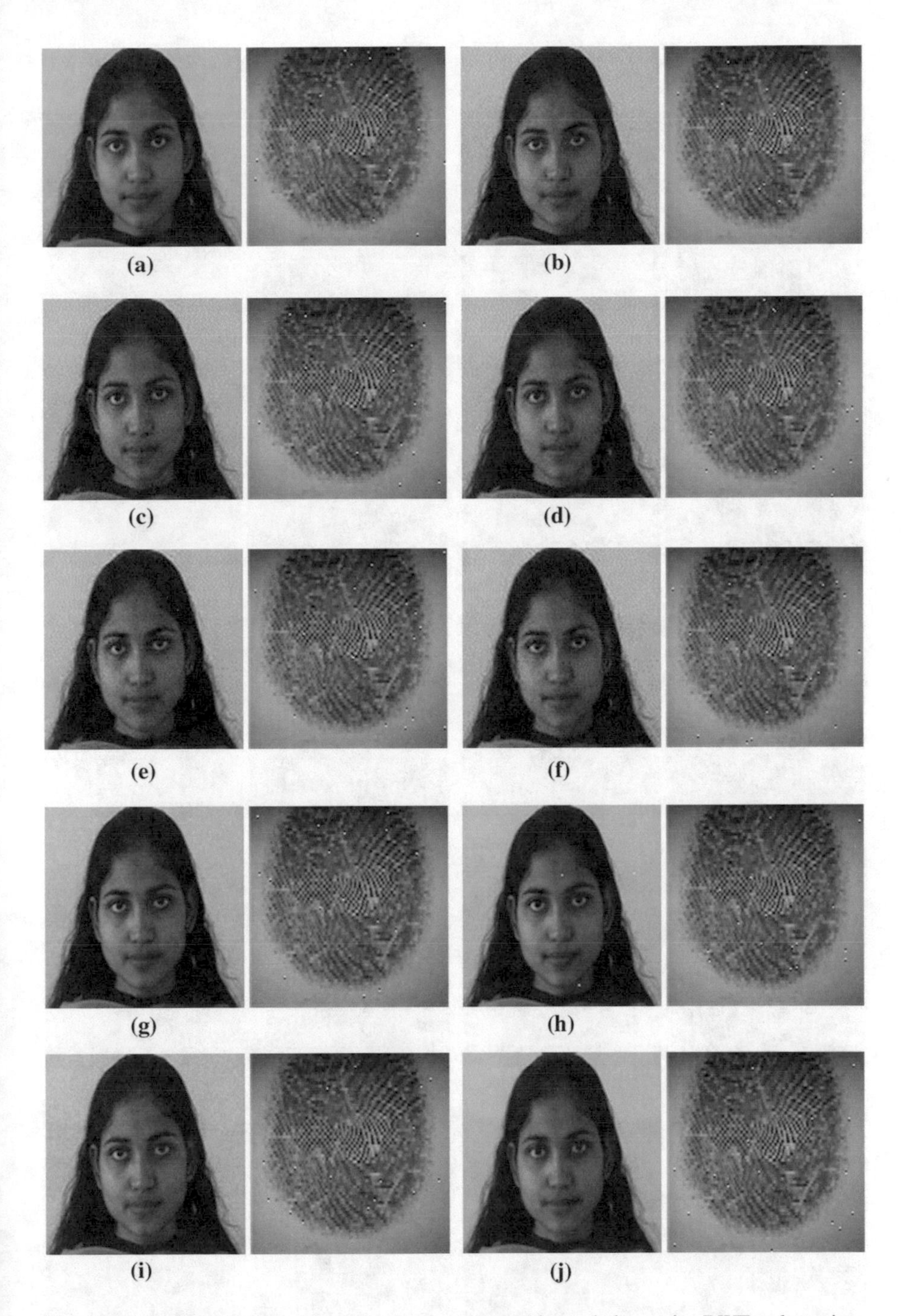

Fig. 4.7 Experimental results of multibiometric watermarking technique using DWT under various watermarking attacks for gain factor $k = 0.2$. (**a**) No attack (**b**) JPEG compression ($Q = 90$). (**c**) JPEG compression ($Q = 80$). (**d**) JPEG compression ($Q = 70$). (**e**) JPEG compression ($Q = 60$). (**f**) JPEG compression ($Q = 50$). (**g**) Gaussian noise attack (**h**) Salt and pepper noise attack .(**i**) Speckle noise attack. (**j**) Median filter attack (3 × 3). (**k**) Median filter attack (5 × 5). (**l**) Median filter attack (7 × 7). (**m**) Mean filter attack (3 × 3). (**n**) Mean filter attack (5 × 5). (**o**) Mean filter attack (7 × 7). (**p**) Gaussian low-pass filter attack (3 × 3). (**q**) Gaussian low-pass filter attack (5 × 5). (**r**) Gaussian low-pass filter attack (7 × 7). (**s**) Sharpening attack. (**t**) Blurring attack. (**u**) Histogram equalization attack. (**v**) Flipping attack. (**w**) Rotation attack (90 degrees). (**x**) Cropping attack (20%)

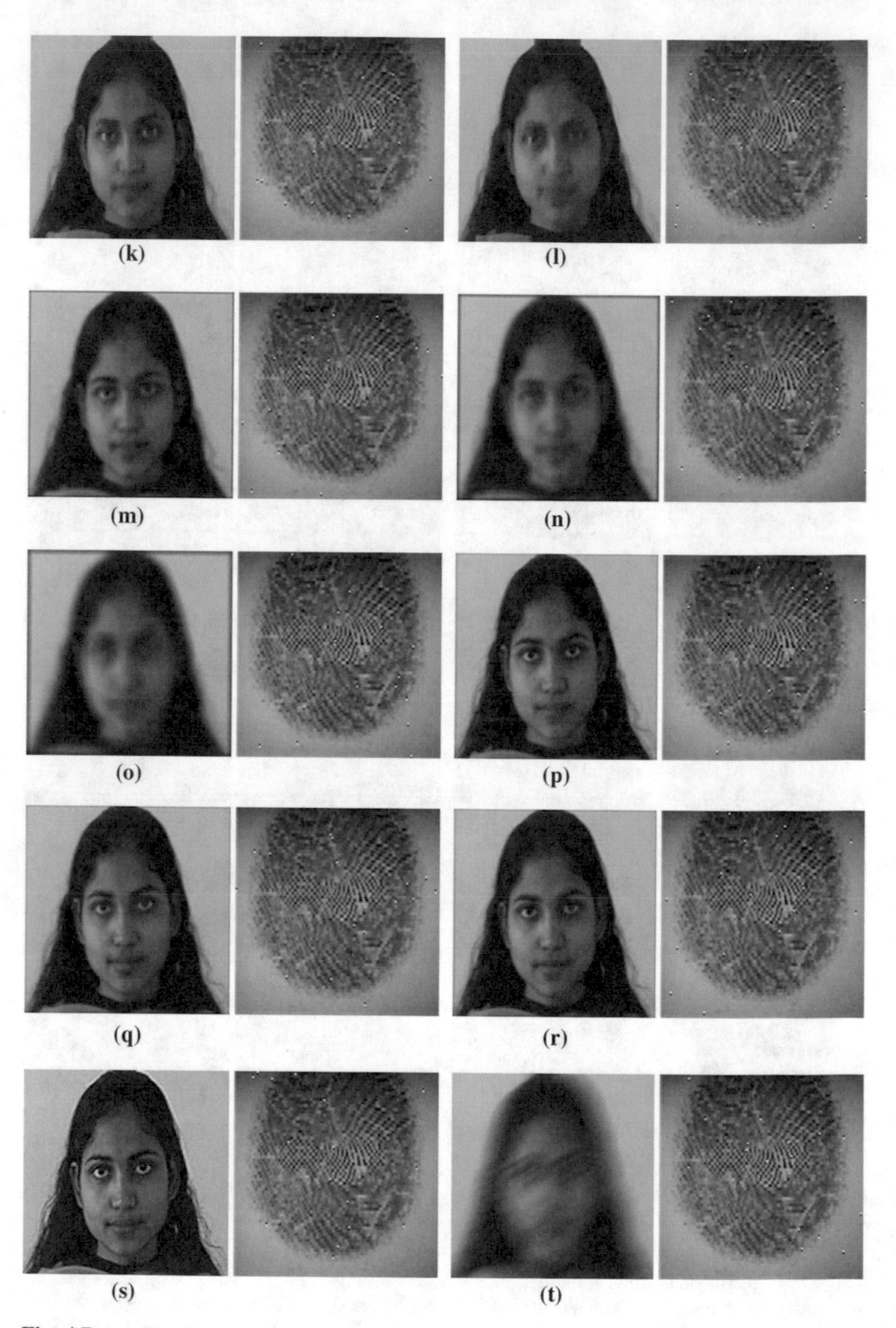

Fig. 4.7 (continued)

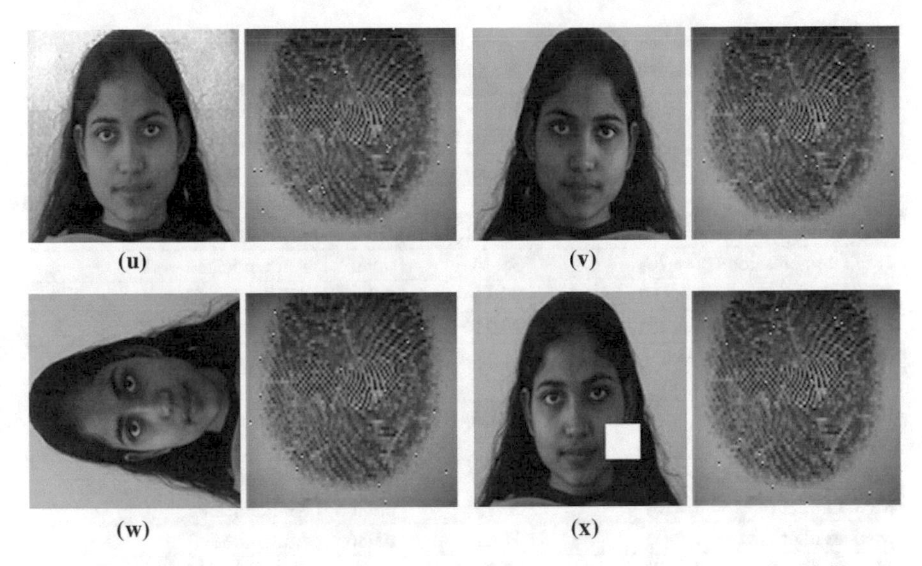

Fig. 4.7 (continued)

SSIM value is above for 0.90 for all watermarking attacks. This situation indicated that this proposed multibiometric watermarking technique using DWT is robust in nature against watermarking attacks.

In the watermark embedding process, the gain factor is multiplied with sparse data of encrypted watermark fingerprint image, and the resultant values are inserted into the wavelet coefficients of host face image. For extraction of encrypted watermark fingerprint image from watermarked face image, gain factor is also required at extraction. Thus, gain factor has significance on PSNR value of watermarked face image as well as SSIM value of extracted watermark fingerprint image. Table 4.2 shows the effect of the gain factor on PSNR value of the watermarked face image and SSIM value of the extracted watermark fingerprint image.

The gain factor value should be 0.1–4.5 for this proposed multibiometric watermarking technique using DWT, because when the gain factor is set to 5 or more, then the PSNR value is less than 28 dB. This situation indicates that gain factor should not be set above 4.5 for this proposed technique. PSNR value below 28 dB is not acceptable for human visual system (HVS) property of watermarking technique.

Table 4.1 PSNR (dB) values and SSIM values for multibiometric watermarking technique using DWT under various watermarking attacks for gain factor $k = 0.2$

Attack	PSNR (dB)	SSIM	Decision about authentication of host face image
No attack	54.96	0.9865	Authenticated
JPEG compression ($Q = 90$)	40.33	0.9866	Authenticated
JPEG compression ($Q = 80$)	37.11	0.9864	Authenticated
JPEG compression ($Q = 70$)	35.44	0.9865	Authenticated
JPEG compression ($Q = 60$)	30.16	0.9865	Authenticated
JPEG compression ($Q = 50$)	29.31	0.9864	Authenticated
Gaussian noise (mean = 0, variance = 0.0001)	38.13	0.9866	Authenticated
Salt and pepper noise (variance = 0.0005)	40.03	0.9866	Authenticated
Speckle noise (variance = 0.0004)	37.97	0.9866	Authenticated
Median filter (size = 3×3)	36.75	0.9866	Authenticated
Median filter (size = 5×5)	31.02	0.9865	Authenticated
Median filter (size = 7×7)	28.28	0.9864	Authenticated
Mean filter (size = 3×3)	25.13	0.9865	Authenticated
Mean filter (size = 5×5)	22.06	0.9865	Authenticated
Mean filter (size = 7×7)	20.25	0.9865	Authenticated
Gaussian low-pass filter (size = 3×3)	34.61	0.9865	Authenticated
Gaussian low-pass filter (size = 5×5)	34.60	0.9865	Authenticated
Gaussian low-pass filter (size = 7×7)	34.57	0.9864	Authenticated
Sharpening	30.64	0.9866	Authenticated
Blurring	19.65	0.9866	Authenticated
Histogram equalization	19.66	0.9866	Authenticated
Flipping	13.90	0.9867	Authenticated
Rotation (90°)	6.90	0.9865	Authenticated
Cropping (20%)	16.17	0.9863	Authenticated

Table 4.2 Effect of gain factor k on performance of multibiometric watermarking technique using DWT

Gain factor	PSNR (dB)	SSIM
0.2	54.96	0.9865
0.5	48.27	0.9866
0.7	44.09	0.9866
1.0	39.96	0.9866
2.0	35.00	0.9865
2.5	33.84	0.9866
3.0	32.48	0.9865
4.0	29.77	0.9865
4.5	29.48	0.9866
5.0	27.89	0.9865

4.4 Analysis of Effect of Proposed Technique Using DWT on Performance of Multibiometric System

In any biometric system, two operations such as verification and authentication of an individual are very important. The performance of any biometric system is measured based on these two operations, so that any biometric image protection technique should not degrade the performance of these two operations of biometric system. Therefore, in this chapter, first, performance analysis of individual watermarked face image-based system and individual extracted watermark fingerprint image-based system is performed. Finally, performance analysis of face-fingerprint-based multibiometric system is performed. This multibiometric system is made by watermarked face-based system and extracted watermark fingerprint-based system.

In this technique, fingerprint information is embedded into the face image. Therefore, in this section, it is checked that insertion of fingerprint information should not change the performance of face-based biometric system. In this section, it is also checked that performance of fingerprint-based biometric system should not change due to CS theory-based encryption process and decryption process.

In order to showcase the effect of watermark fingerprint image on the host face image, face recognition algorithm developed by various researchers (Yang et al. 2000; Lu et al. 2003) is used. In order to showcase the effect of CS theory-based encryption and decryption on watermark fingerprint image, fingerprint recognition algorithm developed by various researchers (Jain et al. 1999; Prabhakar 2001) is used. These two algorithms are selected because the output of the algorithms is given the Euclidean distance between query biometric image and its closest match in the database.

For analysis of verification and authentication performance of the multibiometric system, first, 160 watermarked face images and 160 extracted fingerprint images are stored in the system database of watermarked face-based system and extracted watermark fingerprint-based system, respectively. These watermarked face images and extracted fingerprint images are generated using 50 images from Indian face database, 110 face images from FEI face database, and 80 images from FVC2002 DB3 setB and 80 images from FVC2004 DB4 setB.

For testing of multibiometric system using this proposed technique, 50 images from Indian face database and 110 face images from FEI face database are taken as authentic face images. Also, 50 images from FEI face database and 110 face images from CVL face database are taken as fake face images. These face images are taken as query images for watermarked face-based system. The 80 images from FVC2002 DB3 setB and 80 images from FVC2004 DB4 setB are taken as authentic fingerprint images. Also, 80 images from FVC2002 DB4 setB and 80 images from FVC2004 DB3 setB are taken as fake fingerprint images. These fingerprint images are taken as query images for the extracted watermark fingerprint-based system.

Based on the matching score obtained by recognition algorithms (Yang et al. 2000; Lu et al. 2003; Jain et al. 1999; Prabhakar 2001), the average Euclidean distance for individual watermarked face-based system, individual watermark

Table 4.3 Average Euclidean distance for biometric systems based on proposed technique using DWT (for 160 images)

Average Euclidean distance for watermarked face-based system	Between genuine database and watermarked database	28.09
	Between fake database and watermarked database	505.74
Average Euclidean distance for extracted watermark fingerprint-based system	Between genuine database and extracted database	536.27
	Between fake database and extracted database	711.99
Average Euclidean distance for face-fingerprint-based multibiometric system	Between genuine database and watermarked-extracted database	282.18
	Between fake database and watermarked-extracted database	608.87

fingerprint-based system, and face-fingerprint-based multibiometric system are calculated. The average results for these systems are summarized in Table 4.3.

The threshold distance selected based on this is 450. The average threshold value between a fake biometric database with watermarked face biometric database and extracted fingerprint biometric database is calculated. The average distance value of the impostor biometric database is 608.87. This distance is greater than the selected threshold distance. The average distance value between a genuine face database and genuine fingerprint database with watermarked face biometric database and extracted fingerprint biometric database is also calculated. The average distance value of the genuine biometric database is 282.18. This distance is less than the selected threshold distance, since the average distance between genuine multibiometric database and its watermarked and extracted database is less than the selected threshold distance. This situation is indicated that the performance of biometric database of multibiometric system remains unaffected due to this proposed technique.

The effect of this proposed technique on verification operation and authentication operation of the multibiometric system can be analyzed by various parameters. These parameters such as the probability of verification, False Rejection Rate (FRR), False Acceptance Rate (FAR), and Equal Error Rate (EER) are used for evaluation of the performance of the multibiometric system. The probability of verification of watermarked face-based system and the extracted watermark fingerprint-based system is calculated using Eq. 1.1. The probability of verification of face-fingerprint-based multibiometric system is calculated using Eq. 1.2. The False Rejection Rate (FRR) and False Acceptance Rate (FAR) for various thresholds are calculated using Eqs. 1.3 and 1.4.

Table 4.4 Probability of verification values for biometric systems based on proposed technique using DWT

Threshold	Probability of verification of watermarked face-based system	Probability of verification of extracted watermark fingerprint-based system	Probability of verification of face-fingerprint-based multibiometric system
0.0	0.000	0.000	0.000
0.1	0.013	0.069	0.041
0.2	0.219	0.163	0.191
0.3	0.525	0.294	0.409
0.4	0.625	0.431	0.528
0.5	0.706	0.594	0.650
0.6	0.806	0.763	0.784
0.7	0.931	0.888	0.909
0.8	0.988	0.969	0.978
0.9	1.000	0.994	0.997
1.0	1.000	1.000	1.000

4.4.1 Performance Analysis of Proposed Technique Using DWT for Verification Operation of Multibiometric System

The probability of verification of this proposed technique for face-fingerprint-based multibiometric system for various thresholds is summarized in Table 4.4. The probability of verification curve for this proposed technique for face-fingerprint-based multibiometric system for various thresholds is shown in Fig. 4.8. The verification performance curve for this proposed technique for face-fingerprint-based multibiometric system is shown in Fig. 4.9. It plots the probability of verification versus False Acceptance Rate (FAR). The FAR values for different thresholds are given in Table 4.5. This curve is indicated that this proposed technique using DWT does not degrade the verification performance of the multibiometric system.

4.4.2 Performance Analysis of Proposed Technique Using DWT for Authentication Operation of Multibiometric System

The values of False Rejection Rate (FRR) and False Acceptance Rate (FAR) at a different threshold value for watermarked face-based system, extracted watermark fingerprint-based system, and the face-fingerprint-based multibiometric system are summarized in Table 4.5. Based on values in Table 4.5, plot FRR/FAR vs. threshold curve and receiver operating characteristics (ROC) curve for watermarked face-based system, extracted watermark fingerprint-based system, and the face-

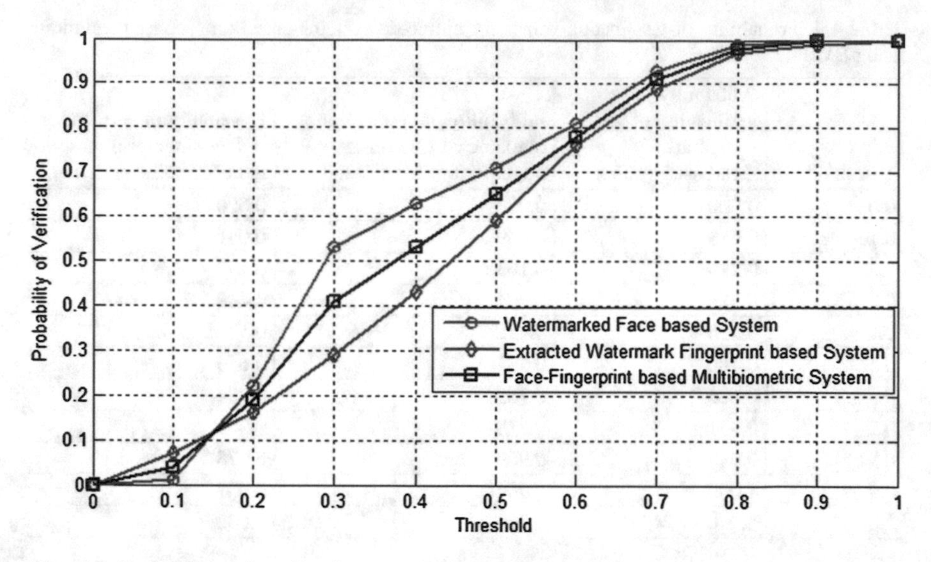

Fig. 4.8 Probability of verification curve for various biometric systems based on proposed technique using DWT

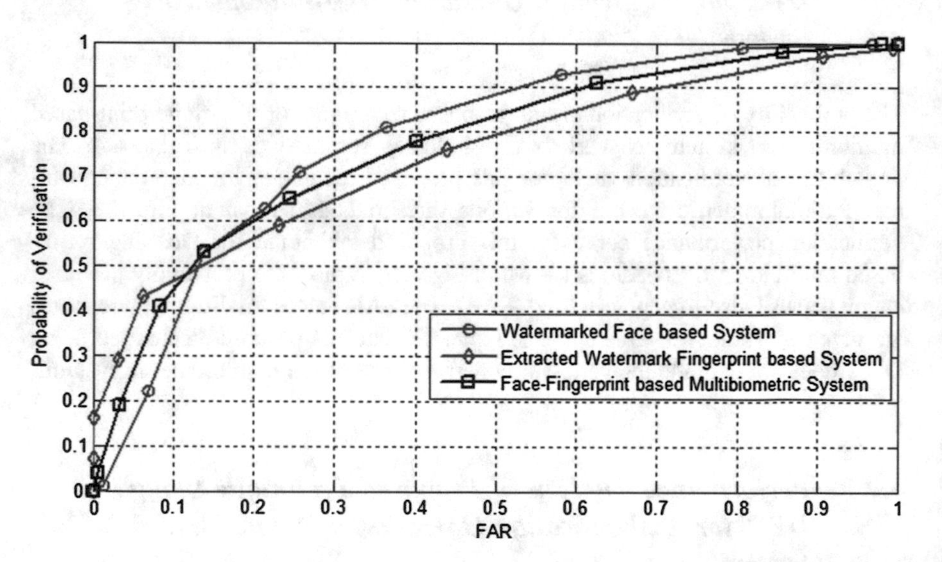

Fig. 4.9 Verification performance curve for various biometric systems based on proposed technique using DWT

fingerprint-based multibiometric system as shown in Figs. 4.10 and 4.11, respectively. Equal Error Rate (EER) is a point on FRR/FAR vs. threshold curve shown in Fig. 4.10 where FAR and FRR have the same value. The EER value of these three biometric systems is summarized in Table 4.6.

Table 4.5 FRR values and FAR values for biometric systems based on proposed technique using DWT

Threshold	FRR of FS	FAR of FS	FRR of FPS	FAR of FPS	FRR of MBS	FAR of MBS
0.0	1.000	0.000	1.000	0.000	1.000	0.000
0.1	0.988	0.013	0.931	0.000	0.959	0.006
0.2	0.781	0.069	0.838	0.000	0.809	0.034
0.3	0.475	0.138	0.706	0.031	0.591	0.084
0.4	0.375	0.213	0.569	0.063	0.472	0.138
0.5	0.294	0.256	0.406	0.231	0.350	0.244
0.6	0.194	0.363	0.238	0.438	0.216	0.400
0.7	0.069	0.581	0.113	0.669	0.091	0.625
0.8	0.013	0.806	0.031	0.906	0.022	0.856
0.9	0.000	0.969	0.006	0.994	0.003	0.981
1.0	0.000	1.000	0.000	1.000	0.000	1.000

FS watermarked face-based system, *FPS* extracted watermark fingerprint-based system, *MBS* face-fingerprint-based multibiometric system

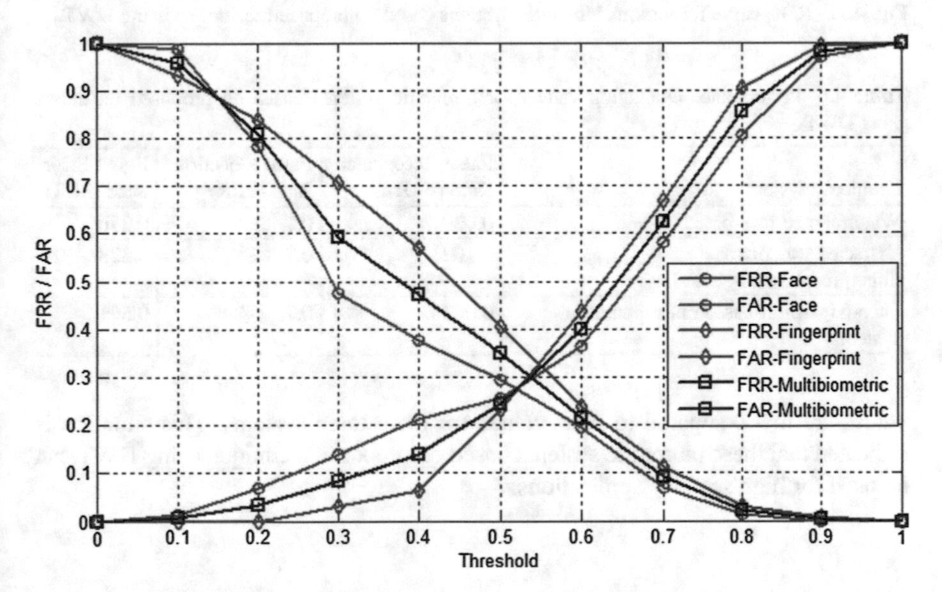

Fig. 4.10 FRR/FAR vs. threshold curve for various biometric systems based on proposed technique using DWT

Based on ROC curve as shown in Fig. 4.11, FRR value 0.7 is chosen as a common value for these three biometric systems, the measured FAR value at that point. The result is summarized in Table 4.6. The results in Table 4.6 show that FAR

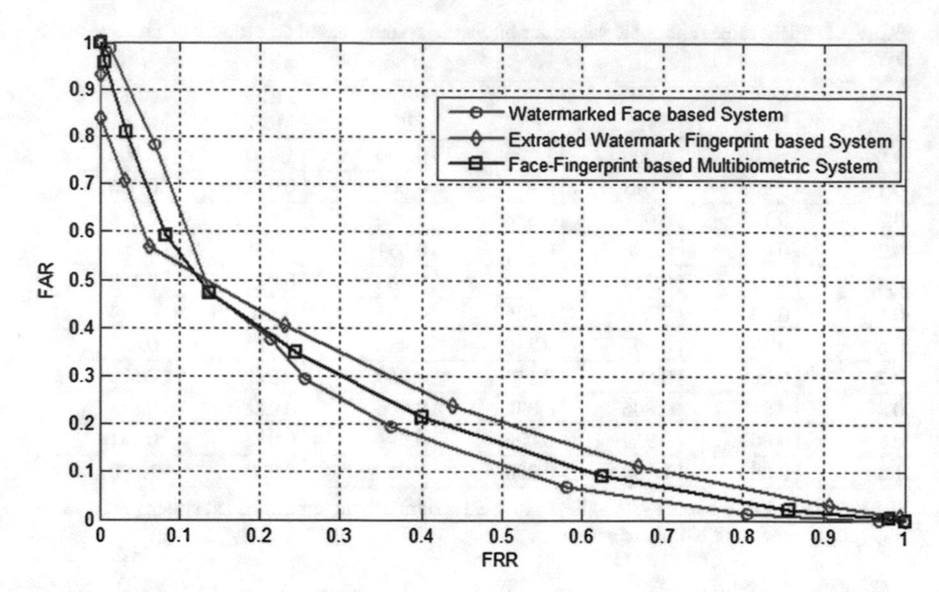

Fig. 4.11 ROC curve for various biometric systems based on proposed technique using DWT

Table 4.6 Performance evaluation values for biometric systems based on proposed technique using DWT

Biometric system	False Acceptance Rate (FAR)	False Rejection Rate (FRR)	Equal Error Rate (EER)
Watermarked face-based system	0.039	0.7	0.276
Extracted watermark fingerprint-based system	0.102	0.7	0.328
Face-fingerprint-based multibiometric system	0.069	0.7	0.301

values are low compared to FRR values for these three systems. This situation is indicated that these biometric systems based on proposed technique using DWT can be used for high security applications.

4.5 Observation of Results

The following observations are made after successfully implementing this proposed watermarking technique for security of biometric image in the multibiometric system.

- This is wavelet-based non-blind watermarking technique. This technique is robust against various watermarking attacks.

- This technique is provided protection to biometric image over a communication channel between two modules of the multibiometric system.
- This tcchniquc is provided security to biometric image against modification attack. It is difficult to generate two biometric images by an impostor, because in this technique, watermark biometric image is encrypted by CS theory, and this encrypted watermark biometric image is embedded into host biometric image.
- This technique does not degrade the verification and authentication performance of the multibiometric system.
- The ROC curve analysis shows that this proposed technique-based multibiometric system can be used in high security applications.
- The limitation of this technique is that approximation wavelet coefficients and correct measurement matrix are required for extraction of watermark fingerprint image at extraction side.

4.6 Comparison of Proposed Technique Using DWT with Existing Techniques Available in Literature

The comparison of proposed technique with existing techniques available in the literature is summarized in Table 4.7. These techniques are compared using various features and parameters. This proposed technique using DWT is embedded sparse data of encrypted biometric data into host biometric image. In the existing watermarking techniques, actual watermark information or noise sequences are embedded into host medium. The existing techniques are provided with computational security using gain factor and particle swarm optimization (PSO) algorithm. This proposed watermarking technique is provided with computational security using CS theory procedure.

The PSNR value in the Rege technique (2014) is 35.18 dB, Kothari technique (2013) is 34.84 dB, Raval technique (2011) is 49.61 dB, Inamdar technique (2010) is 36.32 dB, Jundale technique (2010) is 41.63 dB, Noore technique (2007) is 34.58 dB, and proposed technique is 54.96 dB. The SSIM value in the Rege technique (2014) is 0.977, Kothari technique (2013) is 0.994, Raval technique (2011) is 0.900, Inamdar technique (2010) is 0.891, Jundale technique (2010) is 0.940, Noore technique (2007) is 0.951, and proposed technique is 0.987. These results indicate that performance of the proposed technique is better than existing techniques in terms of imperceptibility, robustness, and security.

Table 4.7 Comparison of proposed watermarking technique using DWT with existing watermarking techniques available in literature

Features and parameters	Rege Tech. et al. (2014)	Kothari Tech. et al. (2013)	Raval Tech. et al. (2011)	Inamdar Tech. et al. (2010)	Jundale Tech. et al. (2010)	Noore Tech. et al. (2007)	Proposed technique
Type of technique	Robust	Robust	Robust	Robust	Robust	Robust	Robust
Used host medium	Standard image	Standard video	Standard image	Standard image	Standard image	Fingerprint image	Face image
Used watermark	Multiple bio-metric features	Standard image	Hash values	Offline signature image	Speech signal	Face image + text information	Sparse information of fingerprint image
Used subband of DWT coefficients	LH of different level	HL of first level	HH of first level	LH, HL, and HH of 2nd level	LH of first level	LH, HL, and HH of second level	LL of fourth level
Security achieved	Gain factor	Gain factor	CS theory	Gain factor	Gain factor	Selected texture region	CS theory
PSNR (dB)	35.18	34.84	49.61	36.32	41.63	34.58	54.96
SSIM	0.977	0.994	0.900	0.891	0.940	0.951	0.987

References

Cox, I., Kilian, J., Shamoon, T., & Leighton, F. (1997). Secure spread spectrum watermarking for multimedia. *IEEE Transactions on Image Processing, 6*(12), 1673–1687.

Gonzales, R., & Woods, R. (2002). *Digital image processing* (pp. 222–226). Upper Saddle River: Prentice Hall, Inc..

Inamdar, V., Rege, P., & Arya, M. (2010). Offline handwritten signature based blind biometric watermarking and authentication technique using biorthogonal wavelet transform. *International Journal of Computer Applications, 11*(1), 19–27.

Jain, A. (1999). *Fundamentals of digital image processing*. Upper Saddle River: Prentice Hall Inc..

Jain, A., Prabhakar, S., & Pankanti, S. (1999). *A Filterbank based representation for classification and matching of fingerprint*. International Joint Conference on Neural Networks (IJCNN), Washington, DC, July, pp. 3284–3285.

Jundale, V., & Patil, S. (2010). Biometric speech watermarking technique in images using wavelet transform. *IOSR Journal of Electronics and Communication Engineering (IOSR-JECE)*, pp. 33–39.

Kothari, A. (2013). *Design, implementation and performance analysis of digital watermarking for video*. Ph.D. thesis, JJTU, India.

Lopez, R., & Boulgouris, N. (2010). *Compressive sensing and combinatorial algorithms for image compression. A project report*. London: King's College London.

Lu, J., Plataniotis, N., & Venetsanopoulos, A. (2003). Face recognition using LDA based algorithms. *IEEE Transactions on Neural Networks, 14*(1), 195–200.

Noore, A., Singh, R., Vatsa, M., & Houck, M. (2007). Enhancing security of fingerprints through contextual biometric watermarking. *Forensic Science International, 169*(2), 188–194.

Prabhakar, S. (2001). *Fingerprint classification and matching using a filterbank*. Ph.D. thesis, Michigan State University, USA.

Raval, M., Joshi, M., Rege, P., & Parulkar, S. (2011). Image tampering detection using compressive sensing based watermarking scheme. *Proceedings of MVIP,* 2011.

Shih, F. (2008). *Digital watermarking and steganography – fundamentals and techniques* (pp. 39–41). Boca Raton: CRC Press.

Vidakovic, B. (1999). *Statistical modelling by wavelets* (pp. 115–116). Wiley .

Yan, J. (2009). *Wavelet matrix*. Victoria: Department of Electrical and Computer Engineering, University of Victoria.

Yang, J., Hua, Y., & William, K. (2000). *An efficient LDA algorithm for face recognition*. Proceedings of the International Conference on Automation, Robotics and Computer Vision (ICARCV 2000), pp. 34–47.

Chapter 5
Multibiometric Watermarking Technique Using Discrete Cosine Transform (DCT) and Discrete Wavelet Transform (DWT)

Abstract This chapter presents technical details and sparsity property of Discrete Cosine Transform (DCT). The hybrid multibiometric watermarking technique using DCT-DWT is explained and analyzed in this chapter. The comparison of presented watermarking technique with existing watermarking techniques is also given in this chapter.

5.1 Discrete Cosine Transform (DCT)

The Discrete Cosine Transform (DCT) is used for converting the image into its frequency coefficients. The advantage of cosine transform is decomposition of the image into the same size into the frequency domain. In many video and image compression algorithms, the DCT is applied to the image for converting into its frequency domain and then performs quantization on these coefficients for data compression (Shih 2008). If $i(x, y)$ is a representation of the image in pixel domain and $I(u, v)$ is a representation of the image in the frequency domain, the general equation for a 2D DCT is (Jain 1999)

$$I(u,v) = \alpha(u) \cdot \alpha(v) \sum_{x=0}^{M-1} \sum_{y=0}^{N-1} i(x,y) \cdot \cos\left[\frac{(2x+1)u\pi}{2M}\right] \cdot \cos\left[\frac{(2y+1)v\pi}{2N}\right] \quad (5.1)$$

where

$$\alpha(u) = \sqrt{\frac{1}{M}} \text{ for, } u = 0; \alpha(u) = \sqrt{\frac{2}{M}} \text{ for, } u = 1, 2, 3, \ldots, M-1$$
$$\alpha(v) = \sqrt{\frac{1}{N}} \text{ for, } v = 0; \alpha(v) = \sqrt{\frac{2}{N}} \text{ for, } v = 1, 2, 3, \ldots, N-1$$

The inverse DCT is calculated using below equation:

© Springer International Publishing AG 2018

R. M. Thanki et al., *Multibiometric Watermarking with Compressive Sensing Theory*, Signals and Communication Technology,
https://doi.org/10.1007/978-3-319-73183-4_5

$$i(x,y) = \sum_{u=0}^{M-1} \sum_{v=0}^{N-1} \alpha(u) \cdot \alpha(v) \cdot \cos\left[\frac{(2x+1)u\pi}{2M}\right] \cdot \cos\left[\frac{(2y+1)v\pi}{2N}\right] \tag{5.2}$$

where $x = 0, 1, 2, \ldots, M–1$ and $y = 0, 1, 2, \ldots, N–1$.

The most convenient method for expressing the 2D DCT is given by matrix production as $I = AiA^T$ and its inverse DCT is $i = A^T IA$, where I and i are image data matrices and A is the DCT basis matrix (Shih 2008). These DCT coefficients indicate the correlation between the original data and its corresponding DCT basis value. These coefficients indicate amplitudes of all cosine waves which are used for reconstruction of the original image in the inverse procedure.

The 2D DCT decomposition of an image in different frequency coefficients is shown in Fig. 5.1. In Fig. 5.1, the black portion shows high AC frequency DCT coefficients, and the white portion shows lowest DC frequency DCT coefficients. The DC coefficients have lower band frequency coefficients which are perfect for watermark embedding but create a problem of perception and vice versa are true for high-frequency AC DCT coefficients. The DCT can be applied to any image using two procedures such as without block procedure and with the block procedure.

The important properties of DCT are given below. The DCT can be used for generation of sparse measurements of any image (Jain 1999).

- The DCT is real and orthogonal.
- The DCT is fast transforming.
- The DCT has an excellent energy compaction for highly correlated data.

When DCT basis matrix is multiplied with an image, then the image is converted into its sparse domain. The DCT basis matrix is generated using the Discrete Cosine

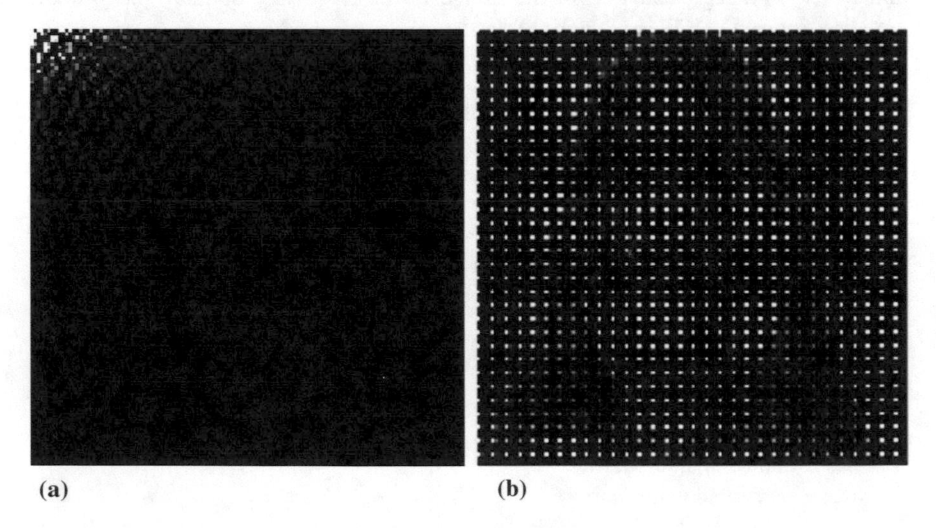

(a) **(b)**

Fig. 5.1 Example of Discrete Cosine Transform (DCT). (**a**) DCT coefficients without block process. (**b**) DCT coefficients with block process

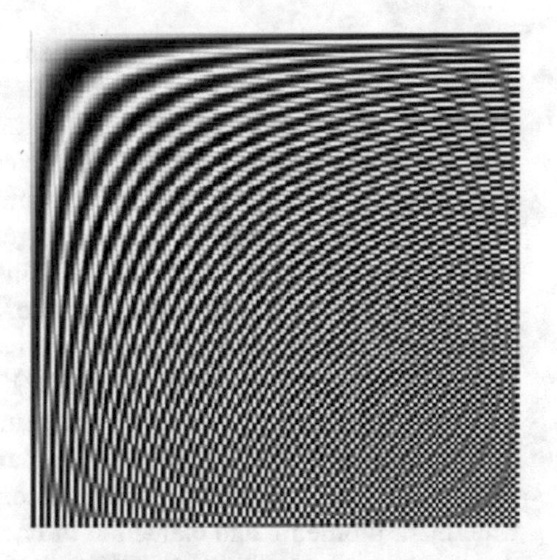

Fig. 5.2 DCT basis matrix

Transform function. This method is described by Tropp and Gilbert (Tropp and Gilbert 2007) and D. Needell (Needell 2009). This method is used for generation of various sizes of DCT basis matrices. The steps for generating of DCT basis matrix in MATLAB are described by Eq. 5.3. The output of this procedure in MATLAB is shown in Fig. 5.2.

$$\begin{aligned}
&\text{for}, x = 1 : 1 : N \\
&\text{for}, y = 1 : 1 : N \\
&\text{DCT_Matrix} = \cos\left(\frac{y*x*\pi}{N}\right) \\
&\text{DCT_Matrix} = \text{DCT_Matrix-mean}(\text{DCT_Matrix}) \\
&\text{DCT_Matrix} = \frac{\text{DCT_Matrix}}{\text{Mean}(\text{DCT_Matrix})} \\
&\text{end} \\
&\text{end}
\end{aligned} \tag{5.3}$$

5.2 Multibiometric Watermarking Technique Using Discrete Cosine Transform (DCT) and Discrete Wavelet Transform (DWT)

In this proposed watermarking technique, watermarking is performed in DCT domain, while encryption of watermark biometric image is performed using DWT basis matrix and CS theory-based encryption process. The DWT basis matrix is used for generation of sparse coefficients of the watermark biometric image. The sparse data of an encrypted watermark biometric image is inserted into DCT coefficients of a host biometric image. This proposed technique is non-blind technique because the original host biometric image is required at the extraction side.

In the presented multibiometric watermarking technique, all DCT coefficients of a host biometric image are modified according to gain factor and values of sparse data of encrypted watermark biometric image. The important data of the image are laid in the low-frequency DCT coefficients which are easily corrupted by any manipulation. When compression is applied to an image which is removed high frequency DCT coefficients of the image. Thus, these two DCT coefficients are easily corrupted by any manipulation. This is the reason behind choosing all DCT coefficients for the watermarking purpose. In this technique, wavelet basis matrix is generated using the wavelet matrix method instead of detail wavelet coefficients (Yan 2009; Vidakovic 1999). This DWT basis matrix with its inverse version is multiplied with the watermark biometric image to convert into its sparse coefficients.

This proposed multibiometric watermarking technique is divided into two procedures such as watermark biometric encryption and embedding of encrypted watermark biometric and extraction of encrypted watermark biometric and decryption of watermark biometric. The proposed block diagram of multibiometric watermarking technique using DCT-DWT is shown in Fig. 5.3.

5.2.1 Watermark Biometric Image Encryption Procedure and Embedding Procedure of Encrypted Watermark Biometric Image

The steps for watermark biometric image encryption procedure and embedding procedure of encrypted watermark biometric image are given below:

- Take a watermark biometric image WBI with a size of $N \times N$ and calculate the size of the image.
- Generate Discrete Wavelet Transform (DWT) basis matrix with a size of $N \times N$ using wavelet basis matrix generation method and Haar wavelet filter.
- The watermark biometric image is converted into its sparse coefficients by multiplying DWT basis matrix with its inverse version with watermark biometric image.

$$x = \Psi_W \times \mathrm{WBI} \times \Psi'_W \tag{5.4}$$

where WBI is a watermark biometric image in terms of the vector, x is sparse coefficients of watermark biometric image, Ψ_W is a DWT basis matrix, and $\Psi_W{}'$ is an inverse DWT basis matrix.

- Generate measurement matrix A with the size of $N \times N$ using a normal distribution with mean $= 0$ and variance $= 1$. This measurement matrix is the same for the embedder side as well as extraction side.
- The CS measurements of a watermark biometric image are generated by multiplying its sparse coefficients with the measurement matrix.

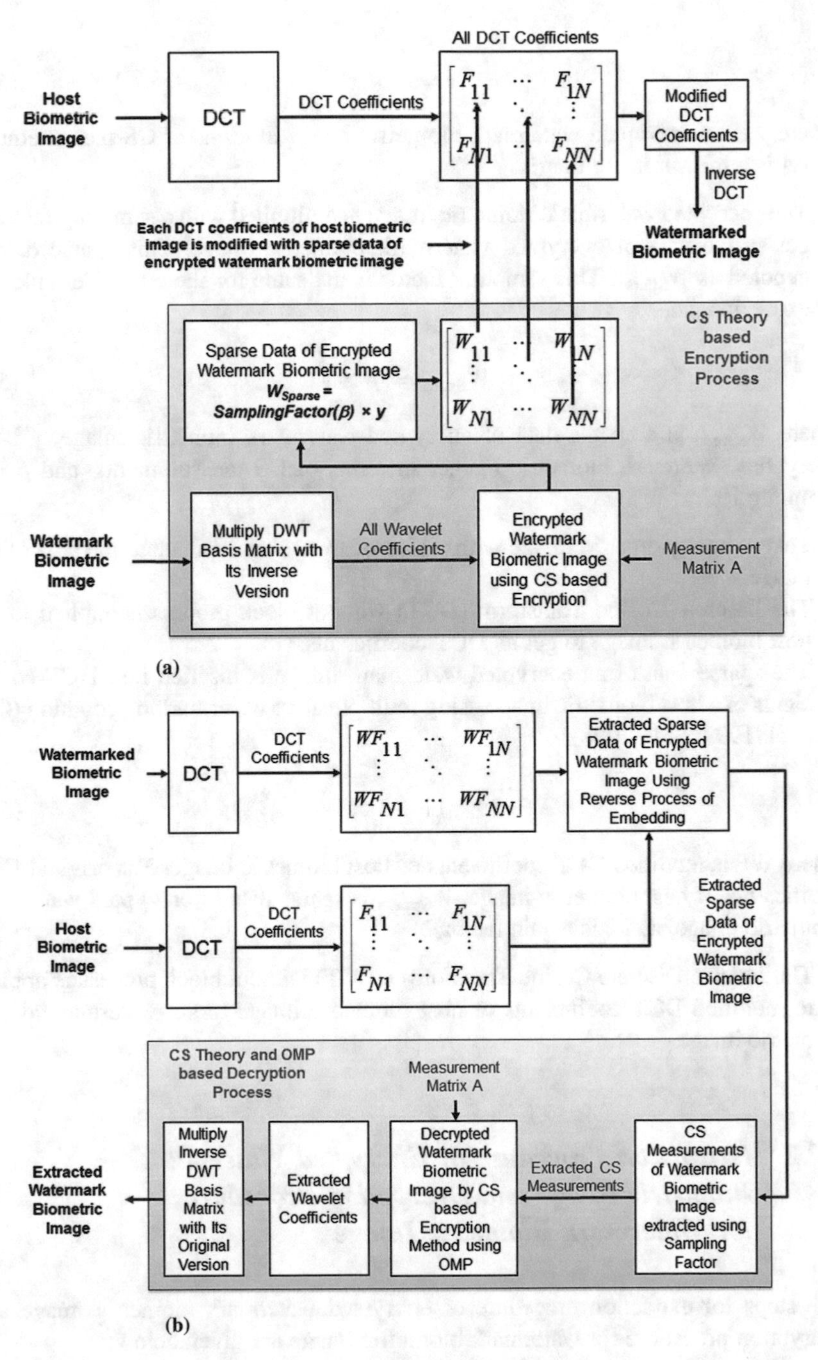

Fig. 5.3 Block diagram of multibiometric watermarking technique using DCT-DWT. (**a**) watermark biometric encryption and embedding of encrypted watermark biometric. (**b**) extraction of encrypted watermark biometric and decryption of watermark biometric

$$y = A \times x \tag{5.5}$$

where y is an encrypted watermark biometric image in terms of CS measurements and A is a measurement matrix.

- The encrypted watermark biometric image is multiplied with a sampling factor to get sparse data of encrypted watermark biometric image. This sparse data is denoted as W_{Sparse}. This sampling factor is the same for the embedder side and extraction side.

$$W_{\text{Sparse}} = \beta \times y \tag{5.6}$$

where W_{Sparse} is a sparse data of encrypted watermark biometric image, y is an encrypted watermark biometric imager in terms of CS measurements, and β is a sampling factor.

- Take a host biometric image with a size of $N \times N$ and calculate the size of the image.
- The Discrete Cosine Transform (DCT) without block process is applied to the host biometric image to get its DCT coefficients.
- The sparse data of an encrypted watermark image is inserted into DCT coefficients of a host biometric image using multiplicative watermarking equation (Cox et al. 1997; Shih 2008).

$$\text{WF} = F * \left(1 + k \times W_{\text{Sparse}}\right) \tag{5.7}$$

where WF is modified DCT coefficients of host biometric image, F is original DCT coefficients of host biometric image, W_{Sparse} is sparse data of encrypted watermark biometric image, and k is a gain factor.

- The Inverse Discrete Cosine Transform (IDCT) without block process is applied to modified DCT coefficients of host biometric image to get watermarked biometric image.

5.2.2 *Extraction Procedure of Encrypted Watermark Biometric Image and Decryption Procedure for Watermark Biometric Image*

The steps for extraction procedure of encrypted watermark biometric image and decryption procedure for watermark biometric image are given below:

- Take a watermarked biometric image which may corrupt or degrade by the impostor. The Discrete Cosine Transform (DCT) without block process is applied to the watermarked biometric image to get its DCT coefficients.

- Take an original host biometric image. The Discrete Cosine Transform (DCT) without block process is applied to the watermarked biometric image to get its DCT coefficients.
- The sparse data of encrypted watermark biometric image is extracted using the reverse procedure of embedding.

$$W_{\text{Extracted}} = \frac{\left(\frac{\text{WF}}{F} - 1\right)}{k} \tag{5.8}$$

where WF is DCT coefficients of watermarked biometric image, F is original DCT coefficients of host biometric image, $W_{\text{Extracted}}$ is extracted sparse data of encrypted watermark biometric image, and k is a gain factor.

- The extracted sparse data of encrypted watermark biometric image is divided by a sampling factor to get extracted encrypted watermark biometric image.

$$y_{\text{Extracted}} = \frac{W_{\text{Extracted}}}{\beta} \tag{5.9}$$

where $W_{\text{Extracted}}$ is extracted sparse data of encrypted watermark biometric image, $y_{\text{Extracted}}$ is extracted encrypted watermark biometric image, and β is a sampling factor.

- After extracting encrypted watermark biometric image, the decryption of watermark biometric image process is performed using CS theory and Orthogonal Matching Pursuit (OMP) algorithm.
- The OMP is applied to extracted encrypted watermark biometric image with correct measurement matrix to get sparse coefficients. These coefficients are all wavelet coefficients of a watermark biometric image.

$$x' = \text{OMP}(y_{\text{Extracted}}, A) \tag{5.10}$$

where x' is an extracted all wavelet coefficient of watermark biometric image.

- Finally, the inverse DWT basis matrix with its original version is multiplied with extracted wavelet coefficients of a watermark biometric image to get extracted watermark biometric image at extraction side.

$$\text{EWBI} = \Psi'_W \times x' \times \Psi_W \tag{5.11}$$

where *EWBI* is an extracted watermark biometric image.

- After extracting watermark biometric image, two hypotheses are formulated for authentication of host biometric image.

- If SSIM (WBI, EWBI) $> \tau$, then host biometric image is authenticated.
- If SSIM (WBI, EWBI) $< \tau$, then host biometric image is unauthenticated.

where τ is a predefined threshold value for a decision about an authenticity of host biometric image.

5.3 Experimental Results

For testing and analysis of this proposed technique, 8-bit grayscale 50 face images from Indian face database and 110 face images from FEI face database are taken as host biometric images. The 8-bit grayscale 80 fingerprint images from FVC2002 DB3 setB and 80 fingerprint images from FVC2004 DB4 setB are taken as watermark biometric images. The size of host face image and watermark fingerprint image is 128 × 128 pixels. The sparse data of encrypted watermark fingerprint image is inserted into host face image using below procedure. First, Discrete Wavelet Transform (DWT) basis matrix is generated with a size of 128 × 128 using Haar wavelet. The DWT basis matrix with its inverse version is multiplied with the watermark fingerprint image to get sparse coefficients x of a watermark fingerprint image with a size of 128 × 128. The measurement matrix A with the size of 128 × 128 is generated using a Gaussian distribution with mean = 0 and variance = 1. The CS measurements of watermark fingerprint image with a size of 128 × 128 are generated using $y_{128 \times 128} = A_{128 \times 128} \times x_{128 \times 128}$. The encrypted watermark fingerprint image is multiplied with sampling factor to get sparse data of encrypted watermark fingerprint image and denoted as W_{Sparse}.

The sparse data W_{Sparse} of encrypted watermark fingerprint image is inserted into DCT coefficients of host face image using gain factor at embedder side. On the extraction side, sparse data $W_{\text{Extracted}}$ of encrypted watermark fingerprint image is extracted using reverse procedure of embedding. The decryption of watermark fingerprint image from extracted encrypted fingerprint image can be done by using below procedure at extraction side.

For decryption of watermark fingerprint image, the input of OMP algorithm is encrypted watermark fingerprint image with a size of 128 × 128 and measurement matrix A with a size of 128 × 128. The output of OMP algorithm is sparse coefficients of watermark fingerprint image with a size of 128 × 128. The inverse DWT basis matrix with its original version is multiplied with these extracted sparse coefficients to get extracted watermark fingerprint image at the extraction side. Figure 5.4 shows original face image and watermarked face image, original fingerprint image and extracted fingerprint image, and encrypted fingerprint image and extracted fingerprint image. These results are generated using gain factor k value 0.2 and sampling factor β value 0.00001.

For testing of nature of proposed watermarking technique, various watermarking attacks such as JPEG compression with different quality factor (Q = 90–50), the addition of noise (Gaussian noise, speckle noise, salt and pepper noise), applied

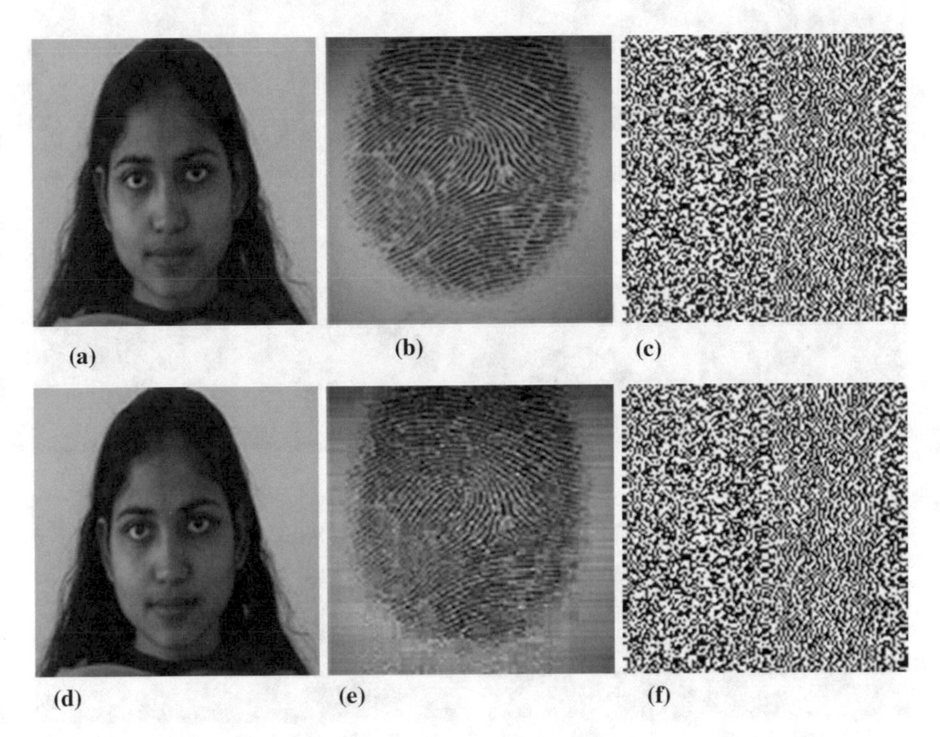

Fig. 5.4 Experimental results of multibiometric watermarking technique using DCT-DWT for gain factor $k = 0.2$. (**a**) Host face image. (**b**) Watermark fingerprint image. (**c**) Encrypted watermark fingerprint image. (**d**) Watermarked face image. (**e**) Extracted watermark fingerprint image. (**f**) Extracted encrypted watermark fingerprint image

different image filter with different filter mask size (median filter, mean filter, and Gaussian low-pass filter), sharpening, blurring, and histogram equalization and different geometric attacks like flipping, rotation, and cropping are applied on watermarked face image. The performance result of watermarking attacks on proposed watermarking technique is shown in Fig. 5.5.

The quality measure such as peak signal to noise ratio (PSNR) is used for quality check between original face and watermarked face image. The structural similarity index measure (SSIM) is used for quality check between the original watermark fingerprint and extracted watermark fingerprint image in proposed multibiometric watermarking technique. The results of multibiometric watermarking technique using DCT-DWT under various watermarking attacks are summarized in Table 5.1.

For authentication of the host face image of an individual, SSIM value between the watermark fingerprint image and extracted watermark fingerprint image must be greater than the predefined threshold value $\tau = 0.90$. SSIM values in Table 5.1 indicate that when a watermarking attack is applied on watermarked face image, the watermark fingerprint image is not extracted successfully at the extraction side and SSIM value is less than 0.90 for all watermarking attacks. This situation indicated

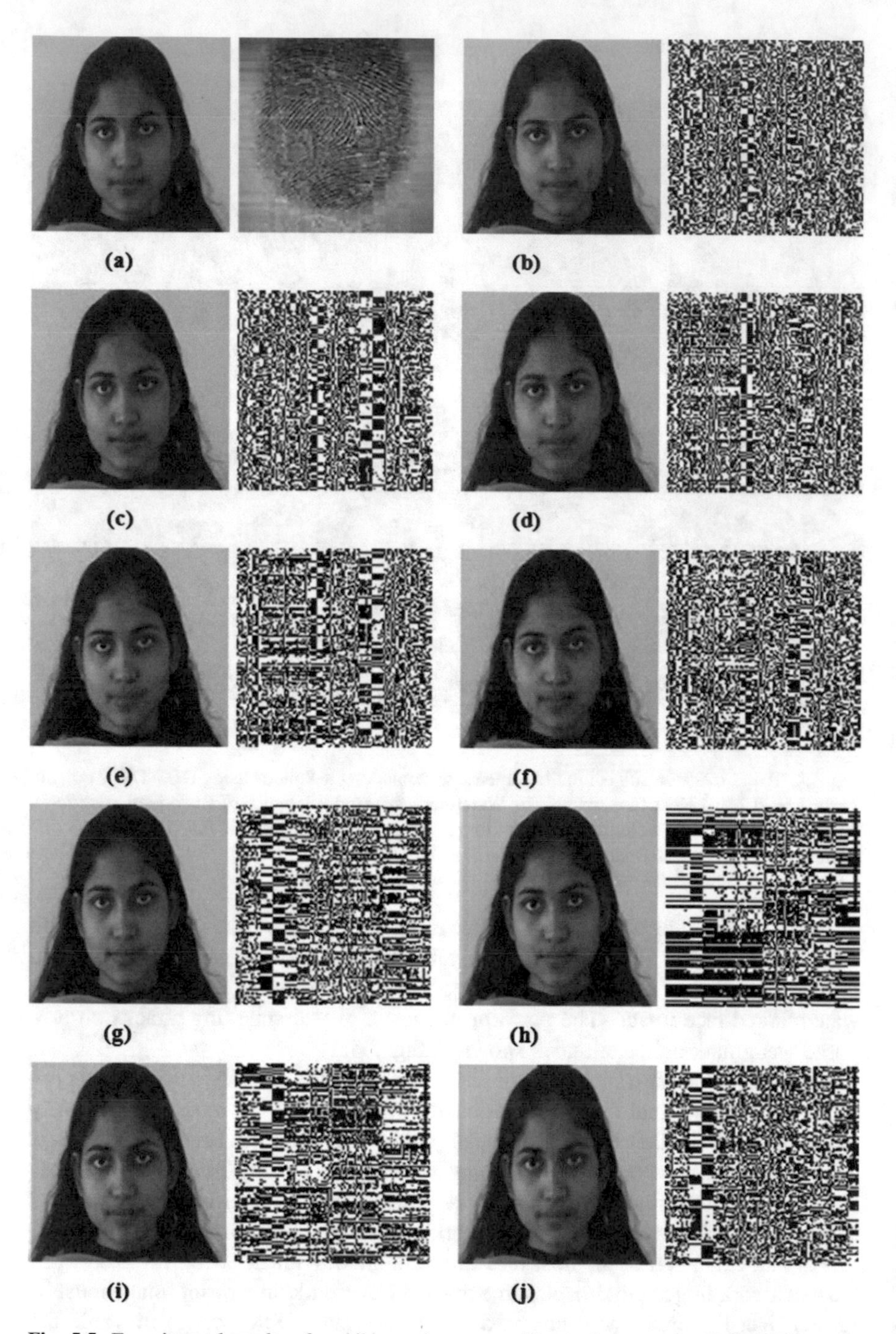

Fig. 5.5 Experimental results of multibiometric watermarking technique using DCT-DWT under various watermarking attacks for gain factor $k = 0.2$. (**a**) No attack. (**b**) JPEG compression ($Q = 90$). (**c**) JPEG compression ($Q = 80$). (**d**) JPEG compression ($Q = 70$). (**e**) JPEG compression ($Q = 60$). (**f**) JPEG compression ($Q = 50$). (**g**) Gaussian noise attack. (**h**) Salt and pepper noise attack. (**i**) Speckle noise attack. (**j**) Median filter attack (3×3). (**k**) Median filter attack (5×5). (**l**) Median filter attack (7×7). (**m**) Mean filter attack (3×3). (**n**) Mean filter attack (5×5). (**o**) Mean filter attack (7×7). (**p**) Gaussian low-pass filter attack (3×3) (**q**) Gaussian low-pass filter attack (5×5). (**r**) Gaussian low-pass filter attack (7×7). (**s**) Sharpening attack. (**t**) Blurring attack. (**u**) Histogram equalization attack. (**v**) Flipping attack. (**w**) Rotation attack (90°). (**x**) Cropping attack (20%)

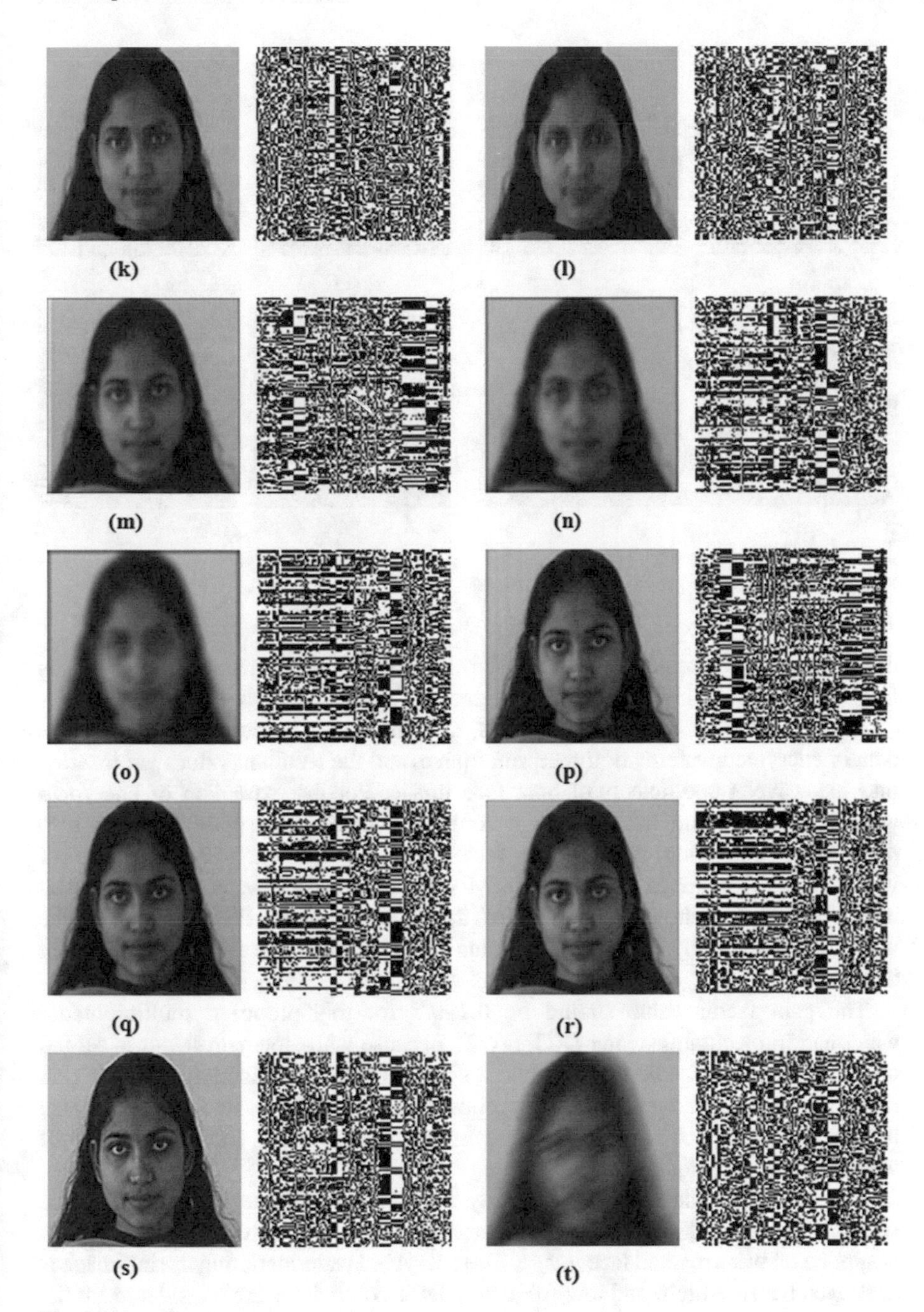

Fig. 5.5 (continued)

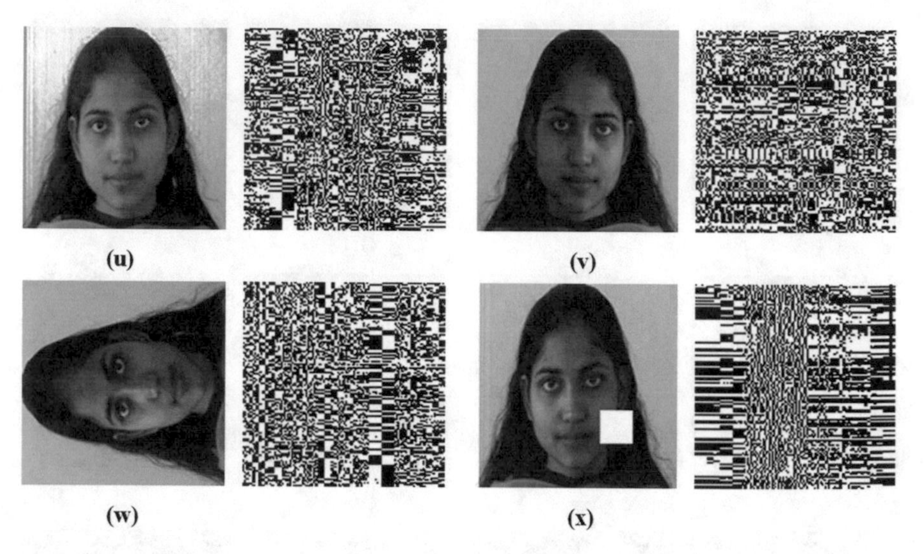

Fig. 5.5 (continued)

that this proposed multibiometric watermarking technique using DCT-DWT is fragile in nature against watermarking attacks.

In the watermark embedding process, the gain factor is multiplied with sparse data of encrypted watermark fingerprint image, and the resultant values are inserted into the wavelet coefficients of host face image. For the extraction of encrypted watermark fingerprint image from watermarked face image, gain factor is also required at extraction. Thus, gain factor has significant on PSNR value of watermarked face image as well as SSIM value of extracted watermark fingerprint image. Table 5.2 shows the effect of the gain factor on PSNR value of the watermarked face image and SSIM value of the extracted watermark fingerprint image.

The gain factor value should be 0.1–0.7 for this proposed multibiometric watermarking technique using DCT-DWT. Because when the gain factor is set 0.8 or more, then PSNR value is less than 28 dB. This situation is indicated that gain factor does not set above 0.8 for this proposed technique. The value of PSNR value below 28 dB is not acceptable for human visual system (HVS) property of watermarking technique.

The reason behind achieving fragility for this proposed technique is that the encrypted watermark fingerprint image are destroyed when a watermarking attack is applied on watermarked face image. The encrypted watermark fingerprint image is destroyed because high- and low-frequency DCT coefficients are considered for the embedding of it. Figure 5.6 shows the histogram of encrypted watermark fingerprint image and extracted encrypted watermark fingerprint image without watermarking attacks and under watermarking attacks. Figure 5.6a indicated the histogram of encrypted watermark fingerprint image, and extracted encrypted watermark

Table 5.1 PSNR (dB) values and SSIM values for multibiometric watermarking technique using DCT-DWT under various watermarking attacks for gain factor $k = 0.2$

Attack	PSNR (dB)	SSIM	Decision about authentication of host face image
No attack	43.62	0.9847	Authenticated
JPEG compression ($Q = 90$)	38.61	0.6735	Unauthenticated
JPEG compression ($Q = 80$)	34.51	0.6785	Unauthenticated
JPEG compression ($Q = 70$)	33.90	0.6727	Unauthenticated
JPEG compression ($Q = 60$)	33.47	0.6691	Unauthenticated
JPEG compression ($Q = 50$)	32.30	0.6741	Unauthenticated
Gaussian noise (mean = 0, variance = 0.0001)	36.27	0.6409	Unauthenticated
Salt and pepper noise (variance = 0.0005)	35.37	0.6702	Unauthenticated
Speckle noise (variance = 0.0004)	29.09	0.6756	Unauthenticated
Median filter (size = 3 × 3)	35.60	0.6711	Unauthenticated
Median filter (size = 5 × 5)	30.70	0.6673	Unauthenticated
Median filter (size = 7 × 7)	27.50	0.6718	Unauthenticated
Mean filter (size = 3 × 3)	25.49	0.6639	Unauthenticated
Mean filter (size = 5 × 5)	22.04	0.6710	Unauthenticated
Mean filter (size = 7 × 7)	20.32	0.6772	Unauthenticated
Gaussian low-pass filter (size = 3 × 3)	31.20	0.6646	Unauthenticated
Gaussian low-pass filter (size = 5 × 5)	28.98	0.6933	Unauthenticated
Gaussian low-pass filter (size = 7 × 7)	23.85	0.6706	Unauthenticated
Sharpening	30.39	0.6766	Unauthenticated
Blurring	19.58	0.6739	Unauthenticated
Histogram equalization	19.32	0.6754	Unauthenticated
Flipping	13.86	0.6682	Unauthenticated
Rotation (90°)	6.90	0.6688	Unauthenticated
Cropping (20%)	15.87	0.6630	Unauthenticated

Table 5.2 Effect of gain factor k on performance of multibiometric watermarking technique using DCT-DWT

Gain factor	PSNR (dB)	SSIM
0.1	50.35	0.9894
0.2	43.62	0.9894
0.3	38.25	0.9891
0.4	36.78	0.9891
0.5	36.35	0.9889
0.6	28.90	0.9891
0.7	28.82	0.9893
0.8	26.42	0.9894
0.9	24.83	0.9890
5.0	27.89	0.9865

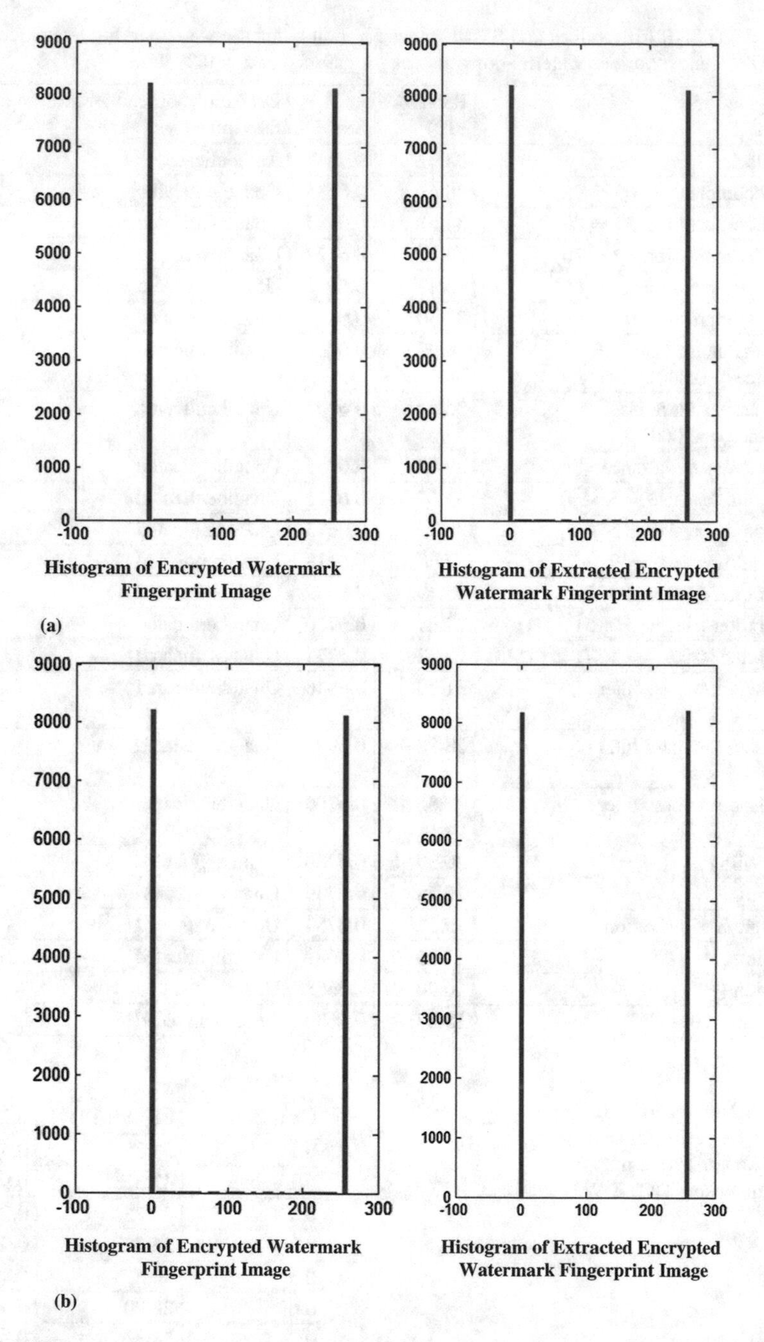

Fig. 5.6 Histogram of encrypted watermark fingerprint image in multibiometric proposed watermarking technique using DCT-DWT. (**a**) Histogram of encrypted watermark fingerprint image without watermarking attack. (**b**) Histogram of encrypted watermark fingerprint image with watermarking attack

fingerprint is the same when a watermarking attack is not applied on watermarked face image. But it is different when a watermarking attack is applied on watermarked face image which is indicated in Fig. 5.6b.

A normalized probability density function (PDF) of encrypted watermark fingerprint image is also calculated for checking fragility of this proposed technique. Figure 5.7 shows normalized PDF of encrypted watermark fingerprint image and extracted encrypted watermark fingerprint image without watermarking attack and under watermarking attack. Figure 5.7a indicated that the normalized PDF of encrypted watermark fingerprint image and extracted encrypted watermark fingerprint image is the same when a watermarking attack is not applied on watermarked face image. But it is different when a watermarking attack is applied on watermarked face image which is indicated in Fig. 5.7b.

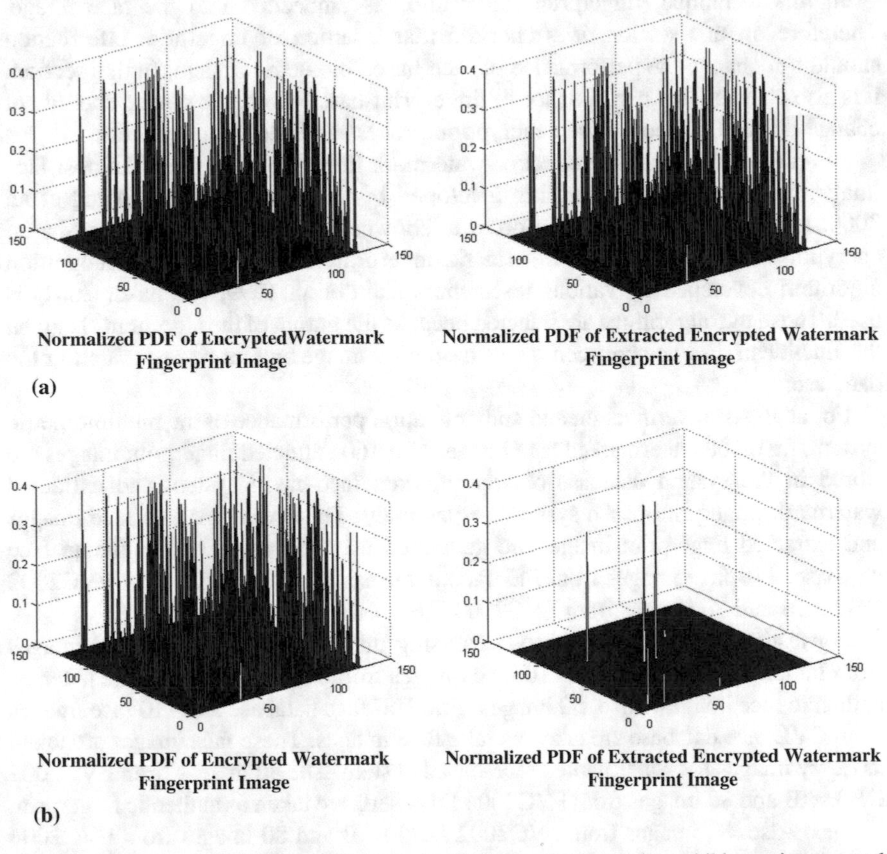

Fig. 5.7 Normalized PDF of encrypted watermark fingerprint image in multibiometric proposed watermarking technique using DCT-DWT. (**a**) Normalized PDF of encrypted watermark fingerprint image without watermarking attack (**b**) Normalized PDF of encrypted watermark fingerprint image with watermarking attack

5.4 Analysis of Effect of Proposed Technique Using DCT-DWT on Performance of Multibiometric System

In any biometric system, two operations such as verification and authentication of an individual are very important. The performance of any biometric system is measured based on these two operations. So that, any biometric image protection technique should not degrade the performance of these two operations of biometric system. Therefore, in this chapter, first, performance analysis of individual watermarked face image-based system and individual extracted watermark fingerprint image-based system is performed. Finally, performance analysis of face-fingerprint-based multibiometric system is performed. This multibiometric system is made by watermarked face-based system and extracted watermark fingerprint-based system.

In this technique, fingerprint information is embedded into the face image. Therefore, in this section, it is checked that insertion of fingerprint information should not change the performance of face-based biometric system. In this section, it is also checked that performance of fingerprint-based biometric system should not change due to CS theory-based encryption process and decryption process.

In order to showcase the effect of watermark fingerprint image on the host face image, face recognition algorithm developed by various researchers (Yang et al. 2000; Lu et al. 2003) is used. In order to showcase the effect of CS theory-based encryption and decryption on watermark fingerprint image, fingerprint recognition algorithm developed by various researchers (Jain et al. 1999; Prabhakar 2001) is used. These two algorithms are selected because the output of the algorithms is given the Euclidean distance between query biometric image and its closest match in the database.

For analysis of verification and authentication performance of the multibiometric system, first, 160 watermarked face images and 160 extracted fingerprint images are stored in the system database of watermarked face-based system and extracted watermark fingerprint-based system, respectively. These watermarked face images and extracted fingerprint images are generated using 50 images from Indian face database, 110 face images from FEI face database, and 80 images from FVC2002 DB3 setB and 80 images from FVC2004 DB4 setB.

For testing of multibiometric system using this proposed technique, 50 images from Indian face database and 110 face images from FEI face database are taken as authentic face images. Also, 50 images from FEI face database and 110 face images from CVL face database are taken as fake face images. These face images are taken as query images for watermarked face-based system. The 80 images from FVC2002 DB3 setB and 80 images from FVC2004 DB4 setB are taken as authentic fingerprint images. Also, 80 images from FVC2002 DB4 setB and 80 images from FVC2004 DB3 setB are taken as fake fingerprint images. These fingerprint images are taken as query images for the extracted watermark fingerprint-based system.

Based on the matching score obtained by recognition algorithms (Yang et al. 2000; Lu et al. 2003; Jain et al. 1999; Prabhakar 2001), the average Euclidean distance for individual watermarked face-based system, individual watermark

Table 5.3 Average Euclidean distance for biometric systems based on proposed technique using DCT-DWT (for 160 images)

Average Euclidean distance for watermarked face-based system	Between genuine database and watermarked database	40.61
	Between fake database and watermarked database	513.13
Average Euclidean distance for extracted watermark fingerprint-based system	Between genuine database and extracted database	616.10
	Between fake database and extracted database	754.20
Average Euclidean distance for face-fingerprint-based multibiometric system	Between genuine database and watermarked-extracted database	328.36
	Between fake database and watermarked-extracted database	633.67

fingerprint-based system, and face-fingerprint-based multibiometric system are calculated. The average results for these systems are summarized in Table 5.3.

The threshold distance selected based on this is 450. The average threshold value between a fake biometric database with watermarked face biometric database and extracted fingerprint biometric database is calculated. The average distance value of the impostor biometric database is 633.67. This distance is greater than the selected threshold distance. The average distance value between a genuine face database and genuine fingerprint database with watermarked face biometric database and extracted fingerprint biometric database is also calculated. The average distance value of the genuine biometric database is 328.36. This distance is less than the selected threshold distance, since the average distance between genuine multibiometric database and its watermarked and extracted database is less than selected threshold distance. This situation is indicated that the performance of biometric database of multibiometric system remains unaffected due to this proposed technique.

The effect of this proposed technique on verification operation and authentication operation of the multibiometric system can be analyzed by various parameters. These parameters such as the probability of verification, False Rejection Rate (FRR), False Acceptance Rate (FAR), and Equal Error Rate (EER) are used for evaluation of the performance of the multibiometric system. The probability of verification of watermarked face-based system and the extracted watermark fingerprint-based system is calculated using Eq. 1.1. The probability of verification of face-fingerprint-based multibiometric system is calculated using Eq. 1.2. The False Rejection Rate (FRR) and False Acceptance Rate (FAR) for various thresholds are calculated using Eqs. 1.3 and 1.4.

Table 5.4 Probability of verification values for biometric systems based on proposed technique using DCT-DWT

Threshold	Probability of verification of watermarked face-based system	Probability of verification of extracted watermark fingerprint-based system	Probability of verification of face-fingerprint-based multibiometric system
0.0	0.000	0.000	0.000
0.1	0.000	0.006	0.003
0.2	0.394	0.044	0.219
0.3	0.625	0.138	0.381
0.4	0.775	0.331	0.553
0.5	0.875	0.481	0.678
0.6	0.950	0.625	0.788
0.7	0.981	0.763	0.872
0.8	0.994	0.944	0.969
0.9	0.994	0.981	0.988
1.0	1.000	1.000	1.000

5.4.1 Performance Analysis of Proposed Technique Using DCT-DWT for Verification Operation of Multibiometric System

The probability of verification of this proposed technique for face-fingerprint-based multibiometric system for various thresholds is summarized in Table 5.4. The probability of verification curve for this proposed technique for face-fingerprint-based multibiometric system for various thresholds is shown in Fig. 5.8. The verification performance curve for this proposed technique for face-fingerprint-based multibiometric system is shown in Fig. 5.9. It plots the probability of verification versus False Acceptance Rate (FAR). The FAR values for different thresholds are given in Table 5.5. This curve is indicated that this proposed technique using DCT-DWT does not degrade the verification performance of the multibiometric system.

5.4.2 Performance Analysis of Proposed Technique Using DCT-DWT for Authentication Operation of Multibiometric System

The values of False Rejection Rate (FRR) and False Acceptance Rate (FAR) at a different threshold value for watermarked face-based system, extracted watermark fingerprint-based system, and the face-fingerprint-based multibiometric system are summarized in Table 5.5. Based on values in Table 5.5, plot FRR/FAR vs. threshold curve and receiver operating characteristics (ROC) curve for watermarked face-

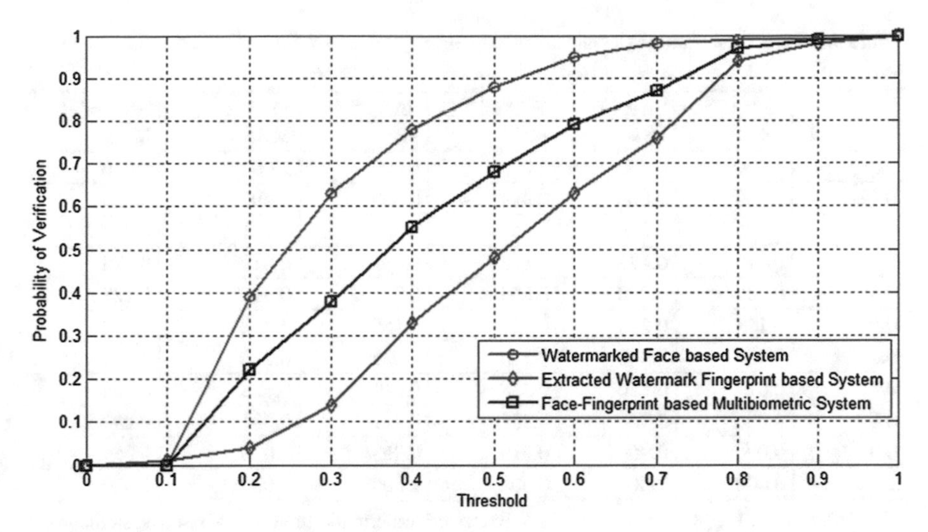

Fig. 5.8 Probability of verification curve for various biometric systems based on proposed technique using DCT-DWT

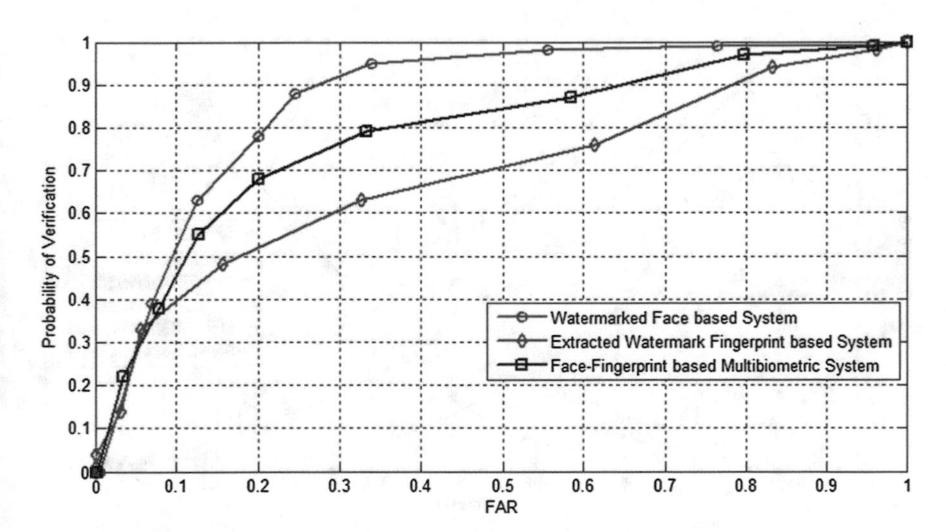

Fig. 5.9 Verification performance curve for various biometric systems based on proposed technique using DCT-DWT

based system, extracted watermark fingerprint-based system, and the face-fingerprint-based multibiometric system is shown in Figs. 5.10 and 5.11, respectively. Equal Error Rate (EER) is a point on FRR/FAR vs. threshold curve shown in Fig. 5.11 where FAR and FRR have the same value. The EER value of these three biometric systems is summarized in Table 5.6.

Table 5.5 FRR values and FAR values for biometric systems based on proposed technique using DCT-DWT

Threshold	FRR of FS	FAR of FS	FRR of FPS	FAR of FPS	FRR of MBS	FAR of MBS
0.0	1.000	0.000	1.000	0.000	1.000	0.000
0.1	1.000	0.006	0.994	0.000	0.997	0.003
0.2	0.606	0.069	0.956	0.000	0.781	0.034
0.3	0.375	0.125	0.863	0.031	0.619	0.078
0.4	0.225	0.200	0.669	0.056	0.447	0.128
0.5	0.125	0.244	0.519	0.156	0.322	0.200
0.6	0.050	0.338	0.375	0.325	0.213	0.331
0.7	0.019	0.556	0.238	0.613	0.128	0.584
0.8	0.006	0.763	0.056	0.831	0.031	0.797
0.9	0.006	0.956	0.019	0.963	0.013	0.959
1.0	0.000	1.000	0.000	1.000	0.000	1.000

FS Watermarked face-based system, *FPS* Extracted watermark fingerprint-based system, *MBS* Face-fingerprint-based multibiometric system

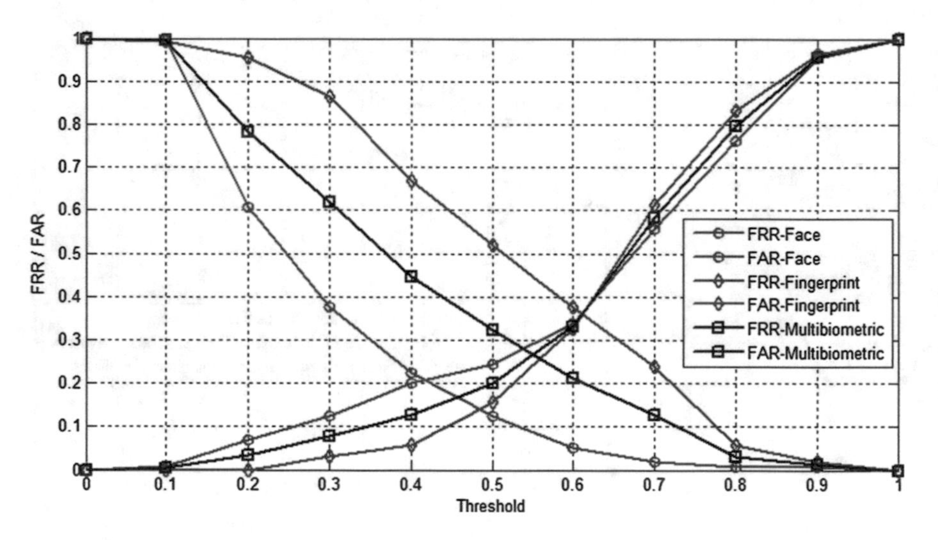

Fig. 5.10 FRR/FAR vs. threshold curve for various biometric systems based on proposed technique using DCT-DWT

Based on ROC curve shown in Fig. 5.11 where FRR value 0.7 is chosen as a common value for these three biometric systems, measure FAR value at that point. The result is summarized in Table 5.6. The results in Table 5.6 show that FAR values are low compared to FRR value for these three systems. This situation is indicated that these biometric systems based on proposed technique using DCT-DWT can be used for high security applications.

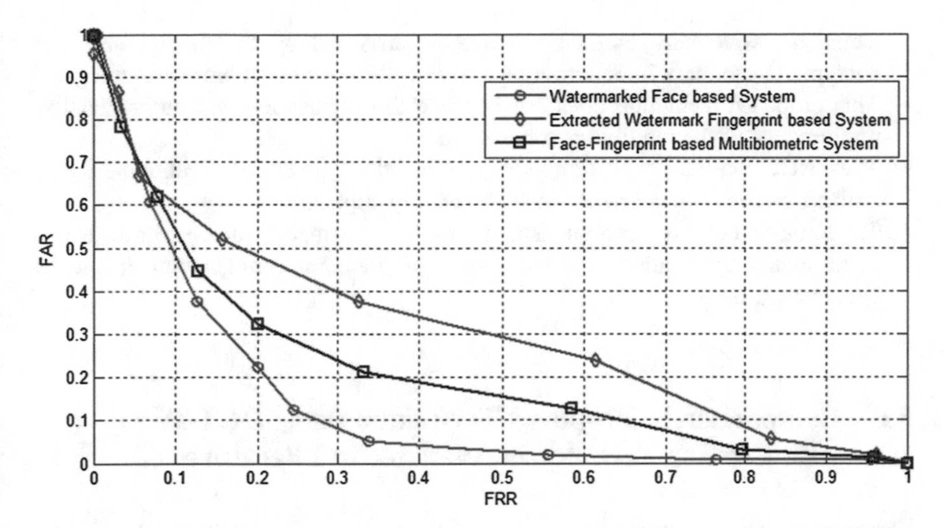

Fig. 5.11 ROC curve for various biometric systems based on proposed technique using DCT-DWT

Table 5.6 Performance evaluation values for biometric systems based on proposed technique using DCT-DWT

Biometric system	False Acceptance Rate (FAR)	False Rejection Rate (FRR)	Equal Error Rate (EER)
Watermarked face-based system	0.010	0.7	0.208
Extracted watermark fingerprint-based system	0.165	0.7	0.359
Face-fingerprint-based multibiometric system	0.075	0.7	0.267

5.5 Observation of Results

The following are some of the observations made after successfully implementing this proposed watermarking technique for security of biometric image in the multibiometric system.

- This is a hybrid non-blind watermarking technique based on Discrete Cosine Transform (DCT) and Discrete Wavelet Transform (DWT). This technique is fragile against various watermarking attacks.
- This proposed technique provides more payload capacity compared to existing watermarking techniques.
- This proposed technique provides authentication to the biometric image in the system database of the multibiometric system.
- This technique provides security to biometric image against modification attack. It is difficult to generate two biometric images by an impostor. Because in this

technique, watermark biometric image is encrypted by CS theory, and this encrypted watermark biometric image is embedded into host biometric image.
- This proposed technique does not degrade the verification and authentication performance of the multibiometric system.
- The ROC curve analysis shows that this proposed technique-based multibiometric system can use in high security applications.
- The limitation of this technique is that wavelet basis matrix and correct measurement matrix is required for extraction of watermark fingerprint image at extraction side.

5.6 Comparison of Proposed Technique Using DCT-DWT with Existing Techniques Available in Literature

The comparison of proposed technique with existing techniques available in the literature is summarized in Table 5.7. These techniques are compared using various features and parameters. This proposed technique using DCT-DWT is embedded sparse data of encrypted biometric data into host biometric image. This proposed watermarking technique is fragile, while existing watermarking techniques in

Table 5.7 Comparison of proposed watermarking technique using DCT-DWT with existing watermarking techniques available in literature

Features and parameters	Paunwala et al. technique (2014)	Bedi et al. technique (2012)	Naik et al. technique (2010)	Rohani et al. technique (2009)	Proposed technique
Type of technique	Robust	Robust	Robust	Robust	Fragile
Used host medium	Standard image	Face image	Fingerprint image	Standard image	Face image
Used watermark	Fusion iris and fingerprint templates	Binary sequence based on fingerprint and demographic data	Face image	Standard image	Sparse measurements of fingerprint image
Used DCT coefficients	Low-frequency AC coefficient of selected DCT block	Find DCT coefficient in block wise using PSO algorithm	First 10 - low-frequency coefficients including the DC coefficient	Find best DCT coefficient in block wise using PSO algorithm	All DCT coefficients
Security achieved	No such scope	PSO algorithm	No such scope	PSO algorithm	CS theory
PSNR (dB)	38.23	40.89	36.66	42.13	43.62
SSIM	–	0.954	0.991	0.987	0.985

literature are robust. This proposed watermarking technique provides computational security using CS theory procedure. The existing watermarking techniques provide computational security using gain factor, PN sequences, and particle swarm optimization (PSO) algorithm.

The PSNR value in the Paunwala technique (2014) is 38.23 dB; in Bedi technique (2012), 40.89 dB; in Naik technique (2010), 36.66 dB; in Rohani technique (2009), 42.13 dB; and in the proposed technique, 43.62 dB. The SSIM value in the Bedi technique (2012) is 0.954; in Naik technique (2010), 0.991; in Rohani technique (2009), 0.987; and in the proposed technique, 0.985. These results indicate that performance of the proposed technique is better than existing techniques in terms of imperceptibility, robustness, and security.

References

Bedi, P., Bansal, R., & Sehgal, P. (2012). Multimodal biometric authentication using PSO based watermarking. *Procedia Technology, 4*, 612–618.

Cox, I., Kilian, J., Shamoon, T., & Leighton, F. (1997). Secure spread spectrum watermarking for multimedia. *IEEE Transactions on Image Processing, 6*(12), 1673–1687.

Jain, A. (1999). *Fundamentals of digital image processing*. Upper Saddle River: Prentice Hall Inc..

Jain, A., Prabhakar, S., & Pankanti, S. (1999). *A Filterbank based representation for classification and matching of fingerprint*. International Joint Conference on Neural Networks (IJCNN), Washington, DC, July, pp. 3284–3285.

Lu, J., Plataniotis, N., & Venetsanopoulos, A. (2003). Face recognition using LDA based algorithms. *IEEE Transactions on Neural Networks, 14*(1), 195–200.

Naik, A., & Holambe, R. (2010). Blind DCT Domain digital watermarking for biometric authentication. *International Journal of Computer Applications (IJCA), 16*(1), 11–15.

Needell, D. (2009). *Topics in compressed sensing*. Ph.D. thesis, University of California, USA.

Paunwala, M., & Patnaik, S. (2014). Biometric template protection with DCT based watermarking. *Machine Vision and Applications, 25*(1), 263–275.

Prabhakar, S. (2001). *Fingerprint classification and matching using a filterbank*. Ph.D. thesis, Michigan State University, USA.

Rohani, M., & Avanaki, A. (2009). A watermarking method based on optimizing SSIM index using PSO in DCT domain. CSICC, pp. 418–423.

Shih, F. (2008). *Digital watermarking and steganography – fundamentals and techniques* (pp. 39–41). Boca Raton: CRC Press.

Tropp, J., & Gilbert, A. (2007). Signal recovery from random measurements via orthogonal matching pursuit. *IEEE Transactions on Information Theory, 53*(12), 4655–4666.

Vidakovic, B. (1999). *Statistical modelling by wavelets* (pp. 115–116). Wiley .

Yan, J. (2009). *Wavelet matrix*. Victoria: Department of Electrical and Computer Engineering, University of Victoria.

Yang, J., Hua, Y., & William, K. (2000). *An efficient LDA algorithm for face recognition*. Proceedings of the International Conference on Automation, Robotics and Computer Vision (ICARCV 2000), pp. 34–47.

Chapter 6
Multibiometric Watermarking Technique Using Discrete Wavelet Transform (DWT) and Singular Value Decomposition (SVD)

Abstract This chapter presents technical details and sparsity property of Singular Value Decomposition (SVD). The hybrid multibiometric watermarking technique using DWT-SVD is explained and analyzed in this chapter. The comparison of presented watermarking technique with existing watermarking techniques is also given in this chapter.

6.1 Singular Value Decomposition (SVD)

An image with size $M \times N$ is represented using Singular Value Decomposition (SVD) into three different matrices which is denoted in Eq. 6.1:

$$[U, S, V] = \mathrm{SVD}(I) \tag{6.1}$$

Singular Value Decomposition (SVD) decomposed the image into three matrices such as a singular value matrix with a size of $M \times N$ and two unitary matrices U with the size of $M \times M$ and V with the size of $N \times N$. The properties of these three matrices (Joshi et al. 2013; Sreedhanya and Soman 2013; Kothari 2013; Jahan 2013; Inamdar and Rege 2012; Mansouri et al. 2009; Ganic and Eskicioglu 2004) are given below.

- It is represented as $I = U{*}S{*}V^T$.
- U is called as a $M \times M$ real or complex unitary matrix, and V^T (the conjugate transpose of V) is called as a $N \times N$ real or complex unitary matrix.
- S is called as a $M \times N$ rectangular diagonal matrix with nonnegative real numbers. This matrix is also called as a singular matrix.
- Brightness and geometric characteristics of the image are represented by this singular matrix.
- This singular matrix is also important for compressive sensing and watermarking because these matrix values are sparse and arranged diagonally.

An image is decomposed into three different matrices after application of SVD on it which is shown in Fig. 6.1. The singular matrix provides sparse property compared

© Springer International Publishing AG 2018

R. M. Thanki et al., *Multibiometric Watermarking with Compressive Sensing Theory*, Signals and Communication Technology,
https://doi.org/10.1007/978-3-319-73183-4_6

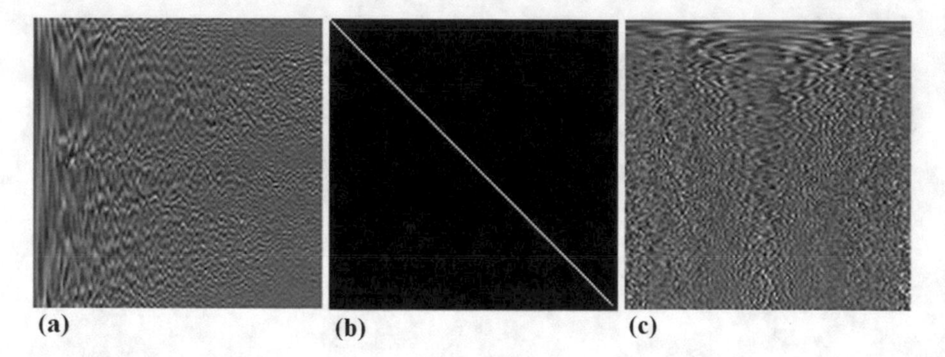

Fig. 6.1 Example of Singular Value Decomposition (SVD). (**a**) U matrix, (**b**) S matrix, (**c**) V matrix

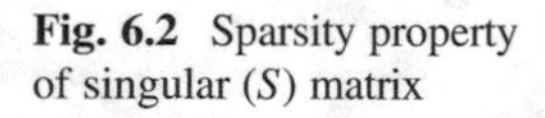

Fig. 6.2 Sparsity property of singular (S) matrix

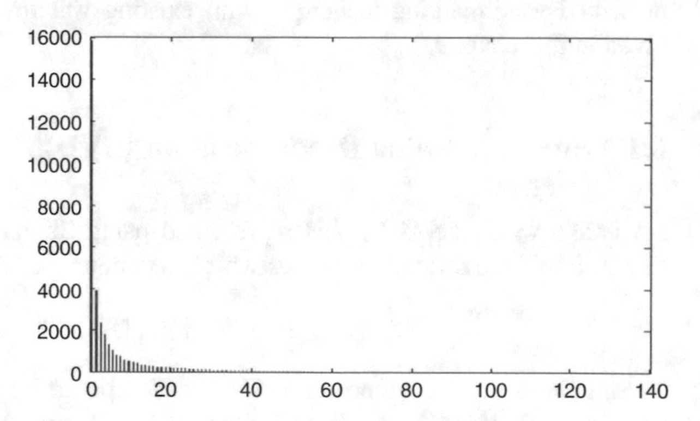

to other two matrices. In the singular matrix, the black portion shows zero values, and the white portion shows nonnegative real values, so that singular matrix values are used as sparse coefficients of an image. The sparsity of singular matrix values is shown in Fig. 6.2 where many singular values are near zero.

6.2 Multibiometric Watermarking Technique Using Discrete Wavelet Transform and Singular Value Decomposition (SVD)

In this proposed watermarking technique, watermarking is performed in DWT + SVD domain, while encryption of watermark biometric image is performed using DWT basis matrix and CS theory-based encryption process. In this proposed watermarking technique, the sparsity property of Singular Value Decomposition (SVD) is explored. In this technique, wavelet basis matrix is generated using the wavelet matrix method instead of detail wavelet coefficients (Yan 2009; Vidakovic 1999).

This DWT basis matrix with its inverse version is multiplied with the watermark biometric image to convert into its wavelet coefficients. The SVD is applied on wavelet coefficients to get U, S, and V matrices of watermark biometric image. Finally, the singular matrix values of watermark biometric image are used as sparse coefficients.

The sparse data of an encrypted watermark biometric image is inserted into the singular value of HL2 wavelet coefficients of a host biometric image. This proposed technique is a non-blind technique because the original host biometric image is required at extraction side. The HL2 wavelet coefficients have low-frequency as well as high-frequency wavelet coefficients. These two coefficients are easily corrupted by any manipulation. This is the reason behind choosing HL2 wavelet coefficients for the watermarking purpose.

This proposed multibiometric watermarking technique is divided into two procedures such as watermark biometric encryption and embedding of encrypted watermark biometric and extraction of encrypted watermark biometric and decryption of watermark biometric. The proposed block diagram of multibiometric watermarking technique using DWT-SVD is shown in Fig. 6.3.

6.2.1 Watermark Biometric Image Encryption Procedure and Embedding Procedure of Encrypted Watermark Biometric Image

The steps for watermark biometric image encryption procedure and embedding procedure of encrypted watermark biometric image are given below:

- Take a watermark biometric image (WBI) with a size of $N \times N$ and calculate the size of the image.
- Generate Discrete Wavelet Transform (DWT) basis matrix with a size of $N \times N$ using wavelet basis matrix generation method and Haar wavelet filter.
- The watermark biometric image is converted into its sparse coefficients by multiplying DWT basis matrix with its inverse version with watermark biometric image.

$$x = \Psi_W \times \text{WBI} \times \Psi_W' \tag{6.2}$$

where WBI is a watermark biometric image in terms of the vector, x is sparse coefficients of watermark biometric image, Ψ_W is a DWT basis matrix, and $\Psi_W{}'$ is an inverse DWT basis matrix.

- A SVD is applied to wavelet coefficients of watermark biometric image to get its singular (S) matrix and orthogonal matrices U and V. The S matrix values of watermark biometric image are taken as sparse coefficients.

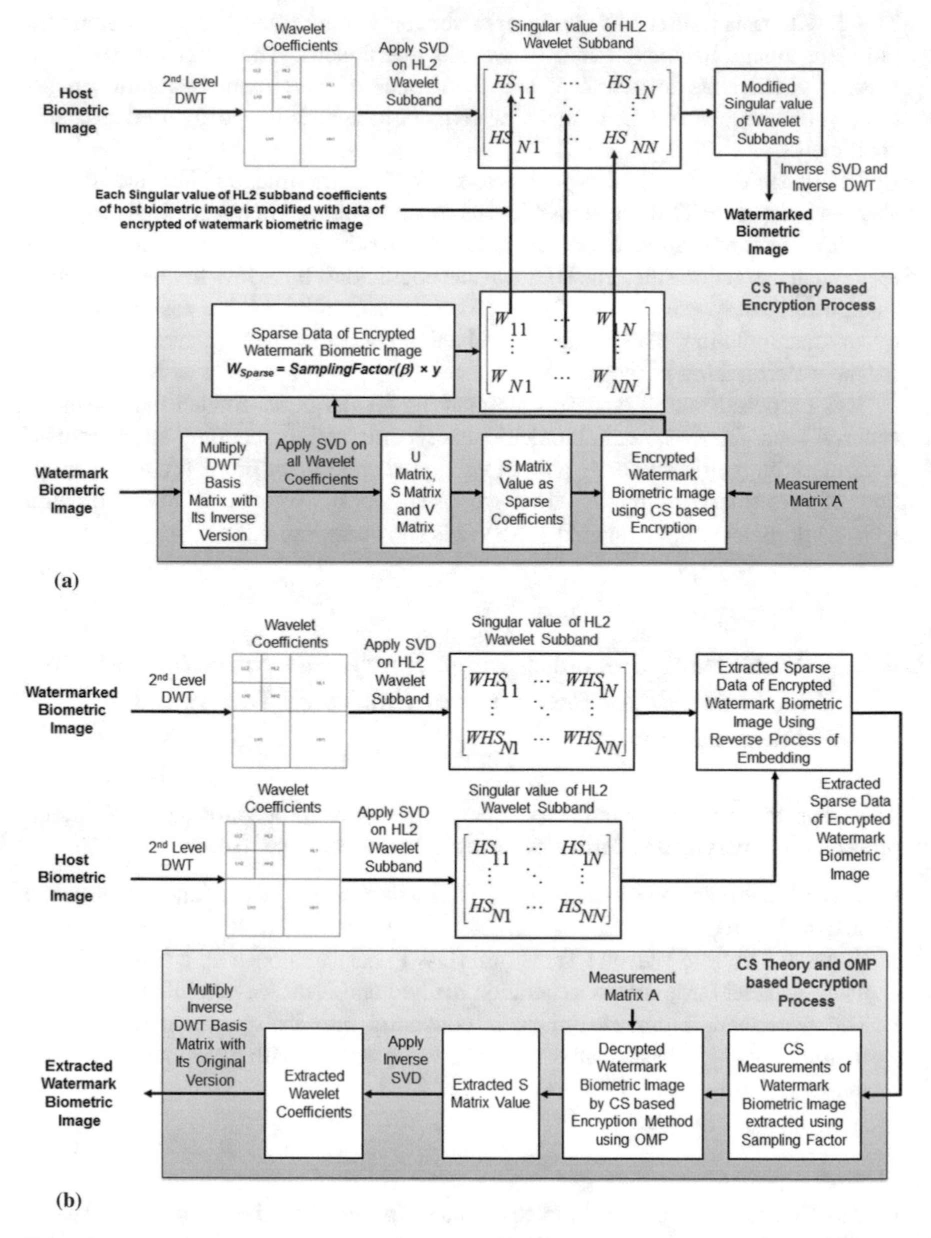

Fig. 6.3 Block diagram of multibiometric watermarking technique using DWT-SVD. (**a**) Watermark biometric encryption and embedding of encrypted watermark biometric. (**b**) Extraction of encrypted watermark biometric and decryption of watermark biometric

$$[U, S, V] = \mathrm{svd}(x) \tag{6.3}$$

- Generate measurement matrix A with the size of $N \times N$ using a normal distribution with mean $= 0$ and variance $= 1$. This measurement matrix is the same for embedder side as well as extraction side.
- The CS measurements of a watermark biometric image are generated by multiplying its sparse coefficients with the measurement matrix.

$$y = A \times S \tag{6.4}$$

where y is an encrypted watermark biometric image in terms of CS measurements and A is a measurement matrix.

- The encrypted watermark biometric image is multiplied with a sampling factor to get sparse data of encrypted watermark biometric image. This sparse data is denoted as W_{Sparse}. This sampling factor is the same for embedder side and extraction side:

$$W_{\mathrm{Sparse}} = \beta \times y \tag{6.5}$$

where W_{Sparse} is a sparse data of encrypted watermark biometric image, y is an encrypted watermark biometric imager in terms of CS measurements, and β is a sampling factor.

- Take a host biometric image with a size of $N \times N$ and calculate the size of the image.
- The second level DWT is applied to the host biometric image to get its wavelet coefficients such as LL2, HL2, LH2, and HH2.
- The Singular Value Decomposition (SVD) is applied on HL2 wavelet subband to get singular value S1 and orthogonal matrices U1 and V1. The S value of HL2 wavelet coefficients of host biometric image is used for watermark embedding.
- The sparse data of an encrypted watermark image is inserted into singular values of HL2 wavelet coefficients of a host biometric image using additive watermarking equation (Cox et al. 1997; Shih 2008).

$$\mathrm{WHS} = \mathrm{HS} + W_{\mathrm{Sparse}} \tag{6.6}$$

where WHS is the modified singular value of HL2 wavelet coefficients of host biometric image, HS is the original singular value of HL2 wavelet coefficients of host biometric image, and W_{Sparse} is the sparse data of encrypted watermark biometric image.

- The inverse SVD is applied to modify singular values matrix with original orthogonal matrices U1 and V1 to get modified HL2 wavelet coefficients.

- The second level DWT is applied to modified HL2 wavelet coefficients with unmodified wavelet coefficients of host biometric image to get watermarked biometric image.

6.2.2 Extraction Procedure of Encrypted Watermark Biometric Image and Decryption Procedure for Watermark Biometric Image

The steps for extraction procedure of encrypted watermark biometric image and decryption procedure for watermark biometric image are given below:

- Take a watermarked biometric image which may be corrupted or degraded by the impostor. The second level DWT is applied to the watermarked biometric image to get its wavelet coefficients such as LLW2, HLW2, LHW2, and HHW2. The SVD is applied to HLW2 to get singular values of watermarked biometric image which is chosen for watermark embedding.
- Take an original host biometric image which may be corrupted or degraded by the impostor. The second level DWT is applied to the watermarked biometric image to get its wavelet coefficients such as LL2, HL2, LH2, and HH2. The SVD is applied to HL2 to get singular values of host biometric image which is chosen for watermark embedding.
- The sparse data of encrypted watermark biometric image is extracted using the reverse procedure of embedding.

$$W_{\text{Extracted}} = \text{WHS} - \text{HS} \tag{6.7}$$

where WHS is the singular value of HL2 wavelet coefficients of watermarked biometric image, HS is the original singular value of HL2 wavelet coefficients of host biometric image, and $W_{\text{Extracted}}$ is the extracted sparse data of encrypted watermark biometric image.

- The extracted sparse data of encrypted watermark biometric image is divided by a sampling factor to get extracted encrypted watermark biometric image.

$$y_{\text{Extracted}} = \frac{W_{\text{Extracted}}}{\beta} \tag{6.8}$$

where $W_{\text{Extracted}}$ is the extracted sparse data of encrypted watermark biometric image, $y_{\text{Extracted}}$ is the extracted encrypted watermark biometric image, and β is a sampling factor.

- After extracting encrypted watermark biometric image, the decryption of watermark biometric image process is performed using CS theory and Orthogonal Matching Pursuit (OMP) algorithm.

- The OMP is applied to extracted encrypted watermark biometric image with correct measurement matrix to get sparse coefficients. These coefficients are singular values of wavelet coefficients of a watermark biometric image.

$$S' = \text{OMP}(y_{\text{Extracted}}, A) \quad (6.9)$$

where S' is the extracted singular values of wavelet coefficient of watermark biometric image.

- The inverse SVD is applied to extracted singular value of wavelet coefficients of watermark biometric image with original orthogonal matrices U and V to get extracted wavelet coefficients of watermark biometric image.

$$x' = U * S' * V^T \quad (6.10)$$

where x' is the extracted wavelet coefficients of watermark biometric image, S' is the extracted singular value of wavelet coefficients of watermark biometric image, and U and V is the original orthogonal matrices.

- Finally, the inverse DWT basis matrix with its original version is multiplied with extracted wavelet coefficients of a watermark biometric image to get extracted watermark biometric image at extraction side.

$$\text{EWBI} = \Psi'_W \times x' \times \Psi_W \quad (6.11)$$

where EWBI is an extracted watermark biometric image.

- After extracting watermark biometric image, two hypotheses are formulated for authentication of host biometric image.
  - If $\text{SSIM}(\text{WBI}, \text{EWBI}) > \tau$ then host biometric image is authenticated.
  - If $\text{SSIM}(\text{WBI}, \text{EWBI}) < \tau$ then host biometric image is unauthenticated.

  where τ is a predefined threshold value used for a decision about an authenticity of host biometric image.

6.3 Experimental Results

For testing and analysis of this proposed technique, 8-bit grayscale 50 face images from Indian face database and 110 face images from FEI face database are taken as host biometric images. The 8-bit grayscale 80 fingerprint images from FVC2002 DB3 setB and 80 fingerprint images from FVC2004 DB4 setB are taken as watermark biometric images. The size of host face image is 512×512 pixels, and the size of watermark fingerprint image is 128×128 pixels. The sparse data of encrypted watermark fingerprint image is inserted into host face image using the below procedure.

First, Discrete Wavelet Transform (DWT) basis matrix is generated with a size of 128 × 128 using Haar wavelet. The DWT basis matrix with its inverse version is multiplied with the watermark fingerprint image to get sparse coefficients of a watermark fingerprint image with a size of 128 × 128. The SVD is applied to wavelet coefficients of a watermark fingerprint image to get U matrix, S matrix, and V matrix and chosen the S matrix values as sparse coefficients of a watermark fingerprint image with a size of 128 × 128. The measurement matrix A with the size of 128 × 128 is generated using a Gaussian distribution with mean = 0 and variance = 1. The CS measurements of watermark fingerprint image with a size of 128 × 128 are generated using $y_{128 \times 128} = A_{128 \times 128} \times x_{128 \times 128}$. The encrypted watermark fingerprint image is multiplied with sampling factor to get sparse data of encrypted watermark fingerprint image and denoted as W_{Sparse}.

The sparse data W_{Sparse} of encrypted watermark fingerprint image is inserted into singular values of HL2 wavelet coefficients of host face image at embedder side. On the extraction side, sparse data $W_{\mathrm{Extracted}}$ of encrypted watermark fingerprint image is extracted using reverse procedure of embedding. The decryption of watermark fingerprint image from extracted encrypted fingerprint image can be done by using the below procedure at extraction side.

For decryption of watermark fingerprint image, the input of OMP algorithm is encrypted watermark fingerprint image with a size of 128×128 and measurement matrix A with a size of 128×128. The output of OMP algorithm is sparse coefficients of watermark fingerprint image with a size of 128×128. The inverse SVD is applied to extracted sparse coefficients with original U matrix and V matrix of watermark biometric image to get wavelet coefficient of watermark biometric image. The DWT basis matrix with its original version is multiplied with these extracted sparse coefficients to get extracted watermark fingerprint image at extraction side.

Figure 6.4 shows original face image and watermarked face image, original fingerprint image and extracted fingerprint image, and encrypted fingerprint image and extracted fingerprint image. These results are generated using sampling factor β value 0.00001.

For testing of the nature of the proposed watermarking technique, various watermarking attacks such as JPEG compression with different quality factor (Q = 90–50), the addition of noise (Gaussian noise, speckle noise, salt and pepper noise), applied different image filter with different filter mask size (median filter, mean filter, and Gaussian low-pass filter), sharpening, blurring, histogram equalization, and different geometric attacks like flipping, rotating, and cropping are applied on watermarked face image. The performance result of watermarking attacks on proposed watermarking technique is shown in Fig. 6.5.

The quality measure such as peak signal to noise ratio (PSNR) is used for quality check between original face and watermarked face image. The structural similarity index measure (SSIM) is used for quality check between the original watermark fingerprint and extracted watermark fingerprint image in proposed multibiometric

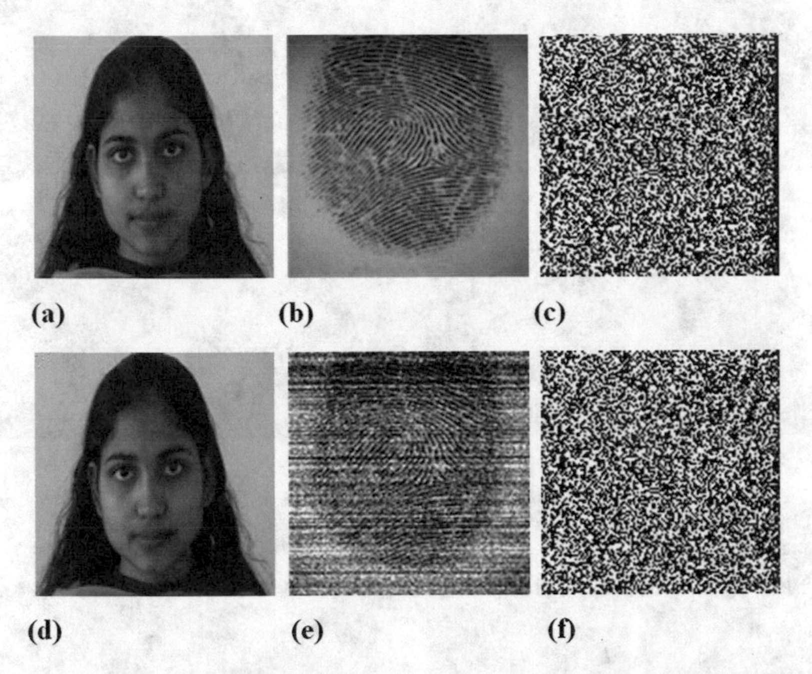

Fig. 6.4 Experimental results of multibiometric watermarking technique using DWT-SVD. (**a**) Host face image. (**b**) Watermark fingerprint image. (**c**) Encrypted watermark fingerprint image. (**d**) Watermarked face image. (**e**) Extracted watermark fingerprint image. (**f**) Extracted encrypted watermark fingerprint image

watermarking technique. The results of multibiometric watermarking technique using DWT-SVD under various watermarking attacks are summarized in Table 6.1.

For authentication of the host face image of an individual, SSIM value between the watermark fingerprint image and extracted watermark fingerprint image must be greater than the predefined threshold value $\tau = 0.90$. SSIM values in Table 6.1 are indicated that when a watermarking attack is applied on watermarked face image, then the watermark fingerprint image is not extracted successfully at extraction side, and SSIM value is less than 0.90 for all watermarking attacks. This situation indicated that this proposed multibiometric watermarking technique using DWT-SVD is fragile in nature against watermarking attacks.

The reason behind achieving fragility for this proposed technique is that the encrypted watermark fingerprint image is destroyed when a watermarking attack is applied on watermarked face image. The encrypted watermark fingerprint image is destroyed because high- and low-frequency wavelet coefficients are considered for embedding of it. Figure 6.6 shows the histogram of encrypted watermark fingerprint image and extracted encrypted watermark fingerprint image without watermarking attacks and under watermarking attacks. Figure 6.6a indicated the histogram of encrypted watermark fingerprint image, and the extracted encrypted watermark fingerprint is the same when a watermarking attack is not applied on watermarked

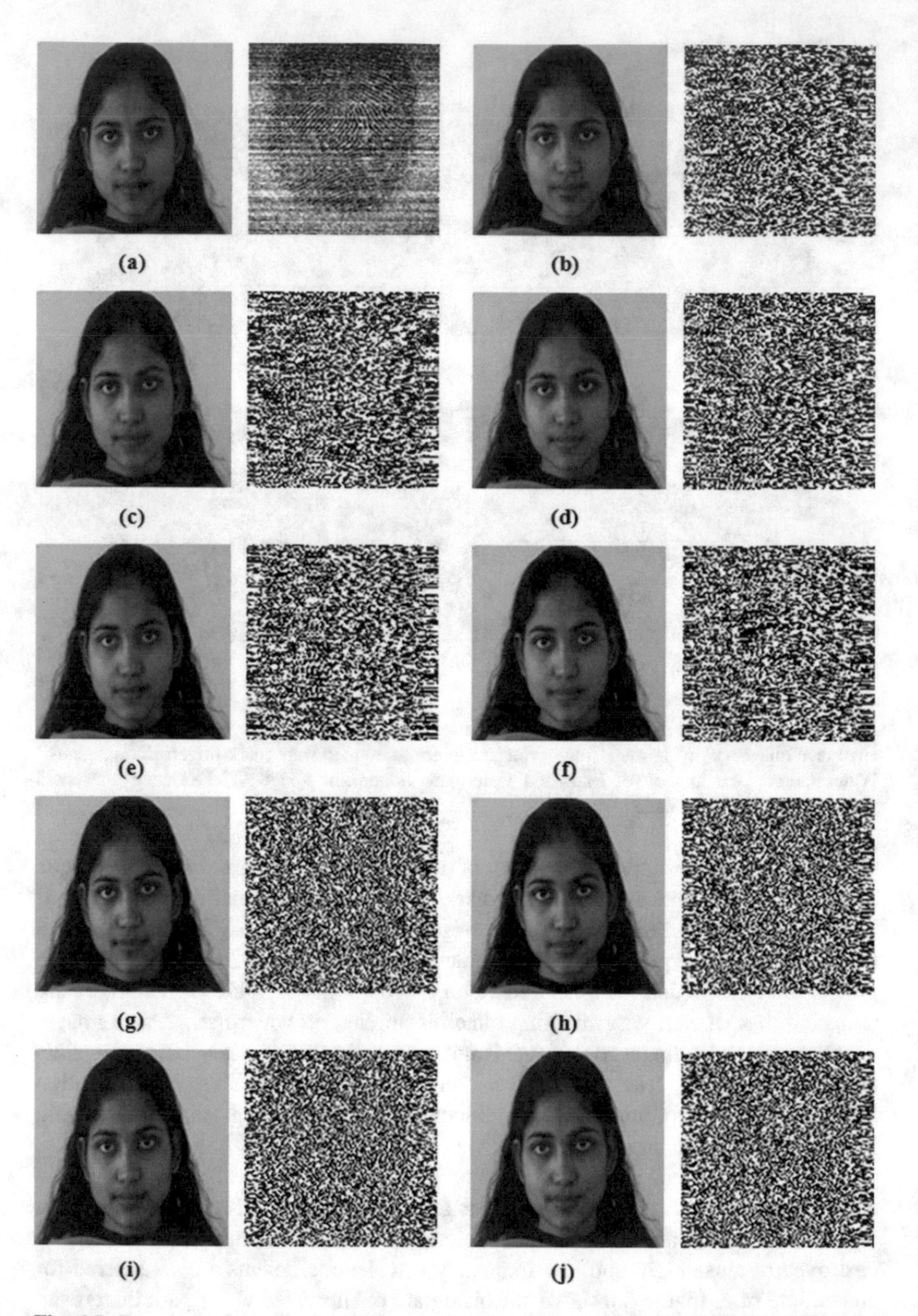

Fig. 6.5 Experimental results of multibiometric watermarking technique using DWT-SVD under various watermarking attacks. (**a**) No attack. (**b**) JPEG compression ($Q = 90$). (**c**) JPEG compression ($Q = 80$). (**d**) JPEG compression. ($Q = 70$). (**e**) JPEG compression ($Q = 60$). (**f**) JPEG compression ($Q = 50$). (**g**) Gaussian noise attack. (**h**) Salt and pepper noise attack. (**i**) Speckle noise attack. (**j**) Median filter attack (3×3). (**k**) Median filter attack (5×5). (**l**) Median filter attack (7×7). (**m**) Mean filter attack (3×3). (**n**) Mean filter attack (5×5) (**o**) Mean filter attack (7×7). (**p**) Gaussian low-pass filter attack (3×3). (**q**) Gaussian low-pass filter attack (5×5). (**r**) Gaussian low-pass filter attack (7×7). (**s**) Sharpening attack. (**t**) Blurring attack. (**u**) Histogram equalization attack. (**v**) Flipping attack. (**w**) Rotation attack (90 degrees). (**x**) Cropping attack (20%)

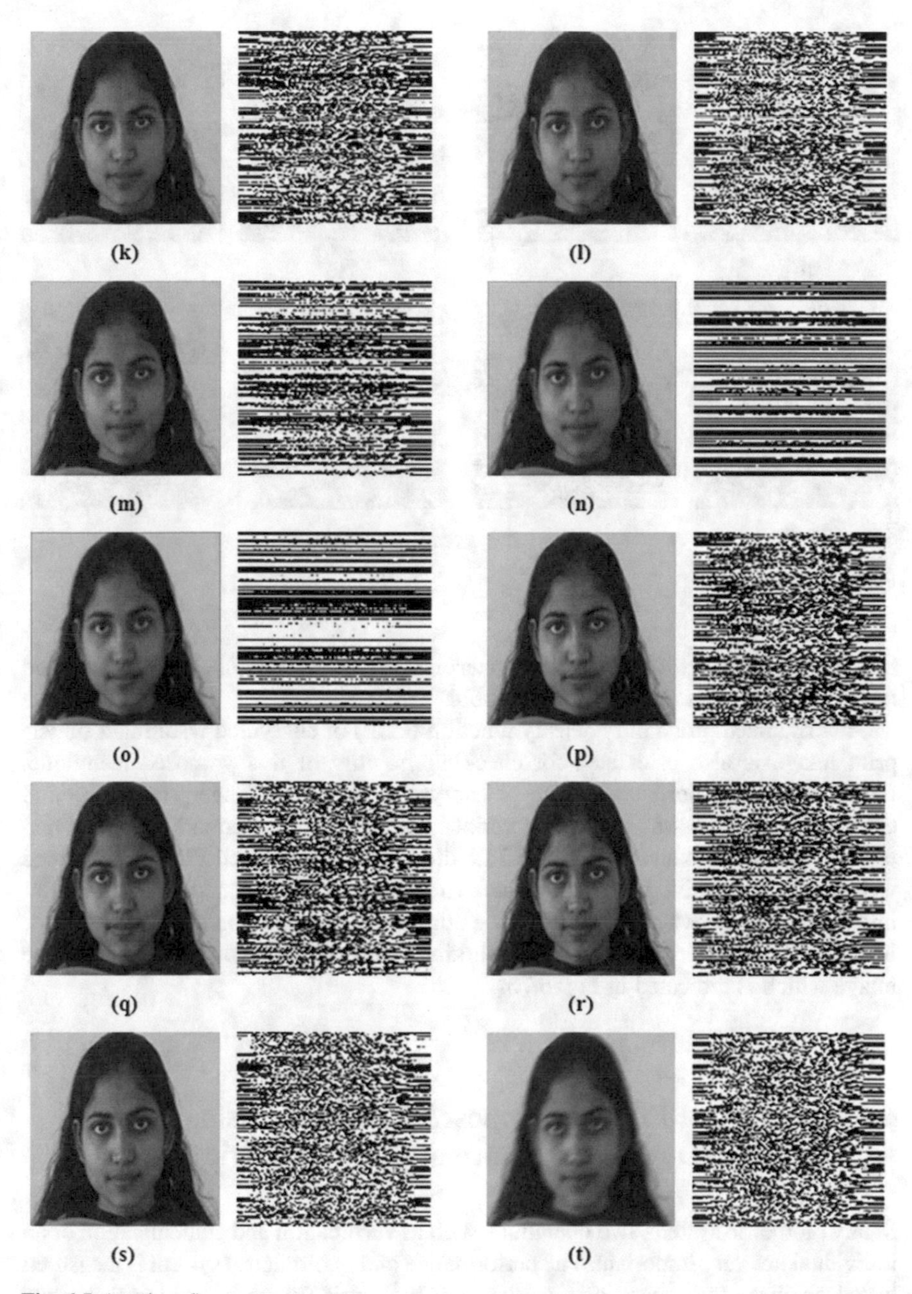

Fig. 6.5 (continued)

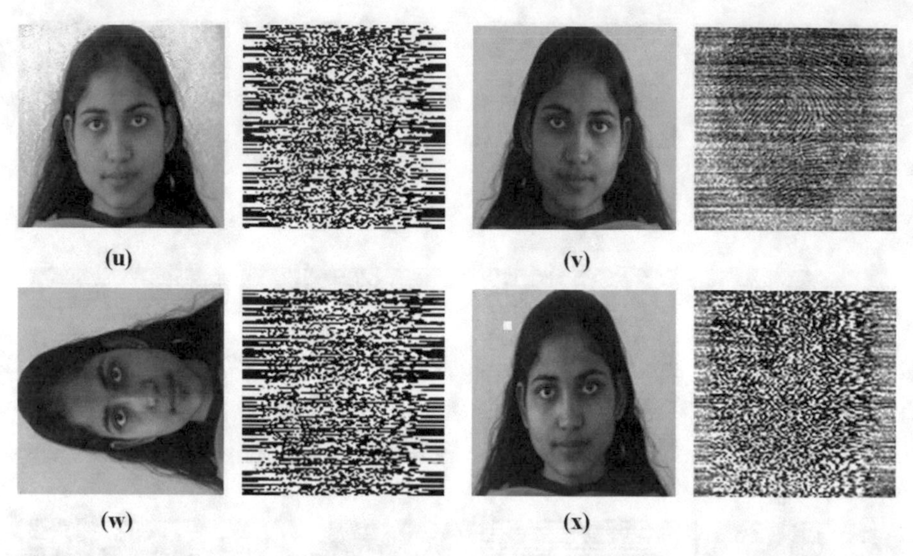

Fig. 6.5 (continued)

face image. But it is different when a watermarking attack is applied on watermarked face image which is indicated in Fig. 6.6b.

A normalized probability density function (PDF) of encrypted watermark fingerprint image is also calculated for checking fragility of this proposed technique. Figure 6.7 shows normalized PDF of encrypted watermark fingerprint image and extracted encrypted watermark fingerprint image without watermarking attack and under watermarking attack. Figure 6.7a indicated the normalized PDF of encrypted watermark fingerprint image, and the extracted encrypted watermark fingerprint image is the same when a watermarking attack is not applied on watermarked face image. But it is different when a watermarking attack is applied on watermarked face image which is indicated in Fig. 6.7b.

6.4 Analysis of Effect of Proposed Technique Using DWT-SVD on Performance of Multibiometric System

In any biometric system, two operations such as verification and authentication of an individual are very important. The performance of any biometric system is measured based on these two operations, so that any biometric image protection technique should not degrade the performance of these two operations of biometric system. Therefore, in this chapter, first, performance analysis of individual watermarked face image-based system and individual extracted watermark fingerprint image-based system is performed. Finally, performance analysis of face-fingerprint-based

Table 6.1 PSNR (dB) values and SSIM values for multibiometric watermarking technique using DWT-SVD under various watermarking attacks

Attack	PSNR (dB)	SSIM	Decision about authentication of host face image
No attack	37.32	0.9504	Authenticated
JPEG compression ($Q = 90$)	36.72	0.3163	Unauthenticated
JPEG compression ($Q = 80$)	36.66	0.1653	Unauthenticated
JPEG compression ($Q = 70$)	36.55	0.1445	Unauthenticated
JPEG compression ($Q = 60$)	36.44	0.1171	Unauthenticated
JPEG compression ($Q = 50$)	36.34	0.0988	Unauthenticated
Gaussian noise (mean = 0, variance = 0.0001)	34.57	0.1611	Unauthenticated
Salt and pepper noise (variance = 0.0005)	33.77	0.1338	Unauthenticated
Speckle noise (variance = 0.0004)	34.51	0.1738	Unauthenticated
Median filter (size = 3 × 3)	37.91	0.1650	Unauthenticated
Median filter (size = 5 × 5)	39.96	0.0129	Unauthenticated
Median filter (size = 7 × 7)	39.66	0.1138	Unauthenticated
Mean filter (size = 3 × 3)	32.19	0.0050	Unauthenticated
Mean filter (size = 5 × 5)	29.75	0.0005	Unauthenticated
Mean filter (size = 7 × 7)	28.00	0.0005	Unauthenticated
Gaussian low-pass filter (size = 3 × 3)	36.94	0.0189	Unauthenticated
Gaussian low-pass filter (size = 5 × 5)	36.95	0.0045	Unauthenticated
Gaussian low-pass filter (size = 7 × 7)	36.95	0.0550	Unauthenticated
Sharpening	32.69	0.0468	Unauthenticated
Blurring	29.12	0.0756	Unauthenticated
Histogram equalization	19.66	0.0077	Unauthenticated
Flipping	14.20	0.9466	Unauthenticated
Rotation (90°)	7.18	0.0016	Unauthenticated
Cropping (20%)	34.77	0.6780	Unauthenticated

multibiometric system is performed. This multibiometric system is made by watermarked face-based system and extracted watermark fingerprint-based system.

In this technique, fingerprint information is embedded into the face image. Therefore, in this section, it is checked that insertion of fingerprint information should not change the performance of face-based biometric system. In this section, it is also checked that performance of fingerprint-based biometric system should not change due to CS theory-based encryption process and decryption process.

In order to showcase the effect of watermark fingerprint image on the host face image, face recognition algorithm developed by various researchers (Yang et al. 2000; Lu et al. 2003) is used. In order to showcase the effect of CS theory-based encryption and decryption on watermark fingerprint image, fingerprint recognition

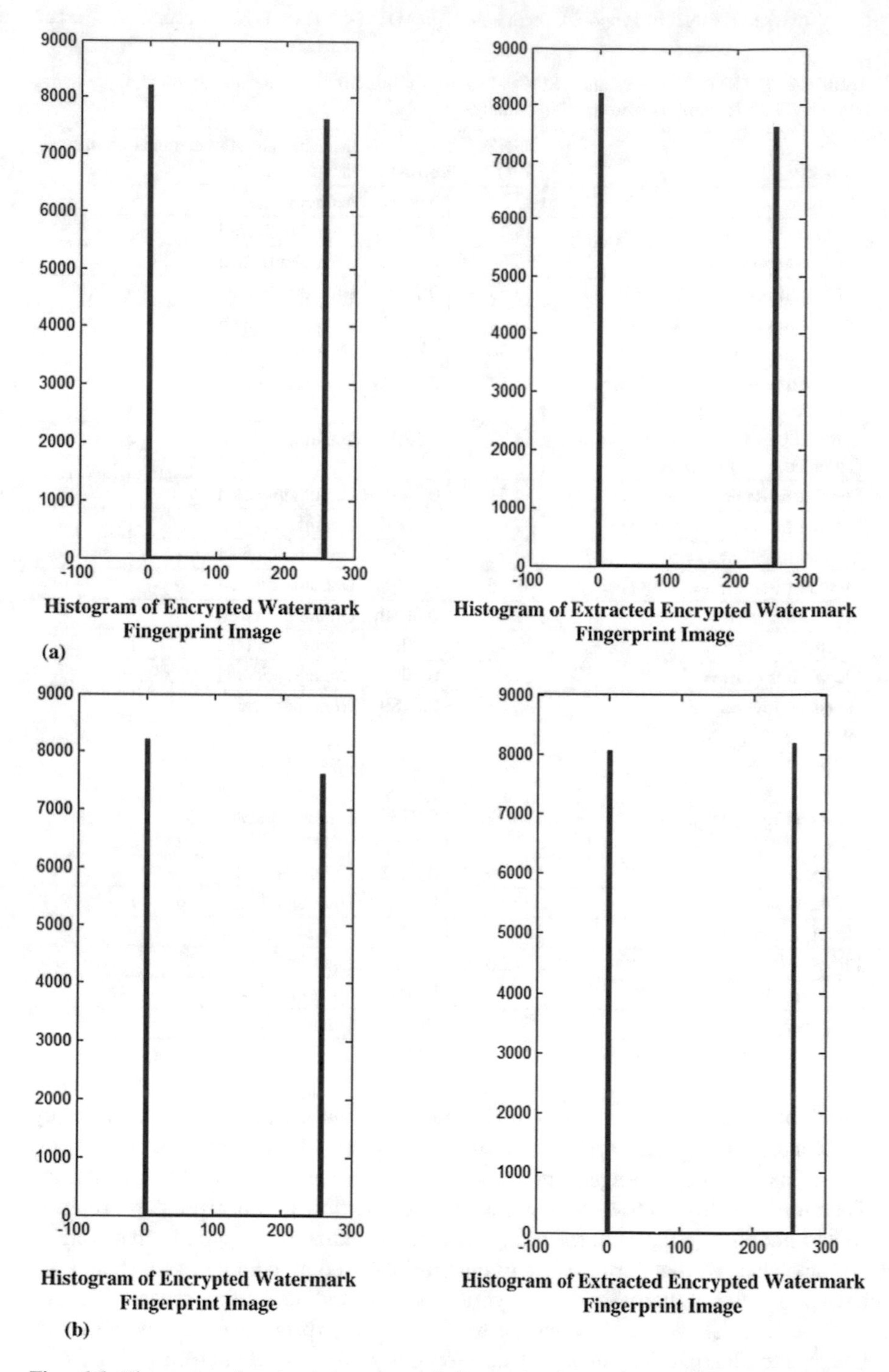

Fig. 6.6 Histogram of encrypted watermark fingerprint image in multibiometric proposed watermarking technique using DWT-SVD. (**a**) Histogram of encrypted watermark fingerprint image without watermarking attack. (**b**) Histogram of encrypted watermark fingerprint image with watermarking attack

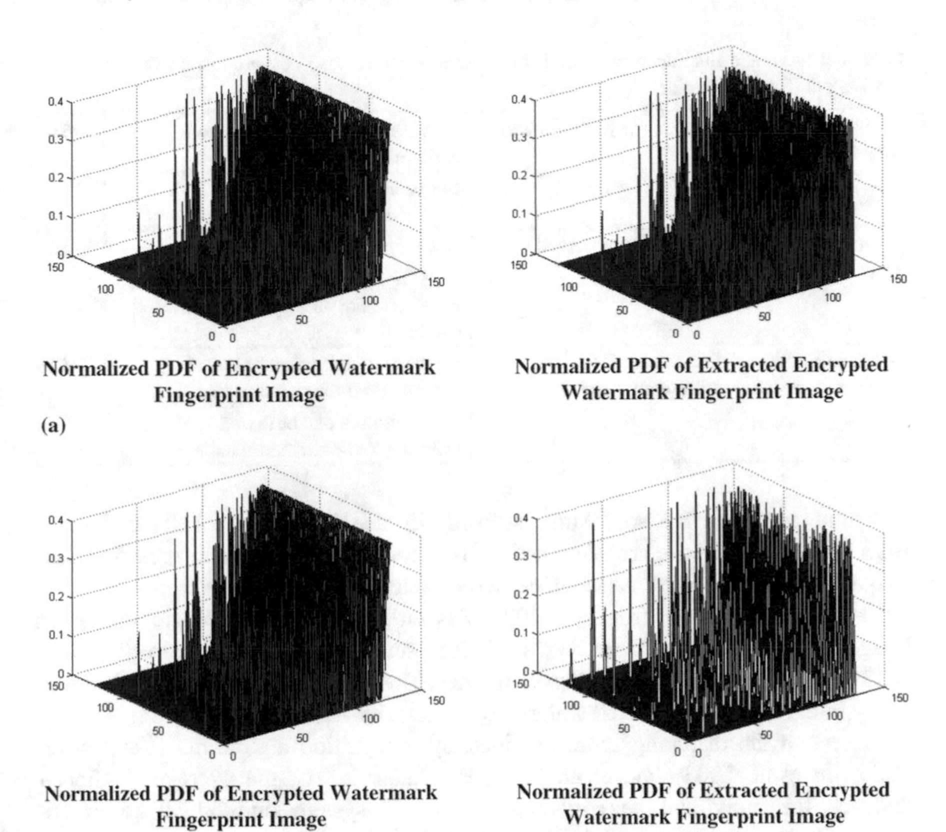

Fig. 6.7 Normalized PDF of encrypted watermark fingerprint image in multibiometric proposed watermarking technique using DWT-SVD. (**a**) Normalized PDF of encrypted watermark fingerprint image without watermarking attack. (**b**) Normalized PDF of encrypted watermark fingerprint image with watermarking attack

algorithm developed by various researchers (Jain et al. 1999; Prabhakar 2001) is used. These two algorithms are selected because the output of the algorithms is given the Euclidean distance between query biometric image and its closest match in the database.

For analysis of verification and authentication performance of the multibiometric system, first, 160 watermarked face images and 160 extracted fingerprint images are stored in the system database of watermarked face-based system and extracted watermark fingerprint-based system, respectively. These watermarked face images and extracted fingerprint images are generated using 50 images from Indian face database, 110 face images from FEI face database, and 80 images from FVC2002 DB3 setB and 80 images from FVC2004 DB4 setB.

For testing of multibiometric system using this proposed technique, 50 images from Indian face database and 110 face images from FEI face database are taken as

Table 6.2 Average Euclidean distance for biometric systems based on proposed technique using DWT-SVD (for 160 images)

Average Euclidean distance for watermarked face-based system	Between genuine database and watermarked database	36.07
	Between fake database and watermarked database	517.08
Average Euclidean distance for extracted watermark fingerprint-based system	Between genuine database and extracted database	502.18
	Between fake database and extracted database	717.93
Average Euclidean distance for face-fingerprint-based multibiometric system	Between genuine database and watermarked-extracted database	269.13
	Between fake database and watermarked-extracted database	617.51

authentic face images. Also, 50 images from FEI face database and 110 face images from CVL face database are taken as fake face images. These face images are taken as query images for watermarked face-based system. The 80 images from FVC2002 DB3 setB and 80 images from FVC2004 DB4 setB are taken as authentic fingerprint images. Also, 80 images from FVC2002 DB4 setB and 80 images from FVC2004 DB3 setB are taken as fake fingerprint images. These fingerprint images are taken as query images for the extracted watermark fingerprint-based system.

Based on the matching score obtained by recognition algorithms (Yang et al. 2000; Lu et al. 2003; Jain et al. 1999; Prabhakar 2001), the average Euclidean distance for individual watermarked face-based system, individual watermark fingerprint-based system, and face-fingerprint-based multibiometric system are calculated. The average results for these systems are summarized in Table 6.2.

The threshold distance selected based on this is 450. The average threshold value between a fake biometric database with watermarked face biometric database and extracted fingerprint biometric database is calculated. The average distance value of the impostor biometric database is 617.51. This distance is greater than the selected threshold distance. The average distance value between a genuine face database and genuine fingerprint database with watermarked face biometric database and extracted fingerprint biometric database is also calculated. The average distance value of the genuine biometric database is 269.13. This distance is less than the selected threshold distance, since the average distance between genuine multibiometric database and its watermarked and extracted database is less than the selected threshold distance. This situation indicated that the performance of biometric database of multibiometric system remains unaffected due to this proposed technique.

The effect of this proposed technique on verification operation and authentication operation of the multibiometric system can be analyzed by various parameters. These parameters such as the probability of verification, False Rejection Rate (FRR), False Acceptance Rate (FAR), and Equal Error Rate (EER) are used for evaluation of the performance of the multibiometric system. The probability of verification of watermarked face-based system and the extracted watermark fingerprint-based system is calculated using Eq. 1.1. The probability of verification of face-fingerprint-based multibiometric system is calculated using Eq. 1.2. The False Rejection Rate (FRR) and False Acceptance Rate (FAR) for various thresholds are calculated using Eqs. 1.3 and 1.4.

6.4.1 Performance Analysis of Proposed Technique Using DWT-SVD for Verification Operation of Multibiometric System

The probability of verification of this proposed technique for face-fingerprint-based multibiometric system for various thresholds is summarized in Table 6.3. The probability of verification curve for this proposed technique for face-fingerprint-based multibiometric system for various thresholds is shown in Fig. 6.8. The verification performance curve for this proposed technique for face-fingerprint-based multibiometric system is shown in Fig. 6.9. It plots the probability of verification versus False Acceptance Rate (FAR). The FAR values for different thresholds are given in Table 6.4. This curve is indicated that this proposed technique using DWT-SVD does not degrade the verification performance of the multibiometric system.

Table 6.3 Probability of verification values for biometric systems based on proposed technique using DWT-SVD

Threshold	Probability of verification of watermarked face-based system	Probability of verification of extracted watermark fingerprint-based system	Probability of verification of face-fingerprint-based multibiometric system
0.0	0.000	0.006	0.003
0.1	0.006	0.038	0.022
0.2	0.025	0.200	0.113
0.3	0.106	0.388	0.247
0.4	0.388	0.581	0.484
0.5	0.768	0.756	0.762
0.6	0.925	0.850	0.888
0.7	0.981	0.950	0.966
0.8	0.994	0.969	0.981
0.9	0.994	0.994	0.994
1.0	1.000	1.000	1.000

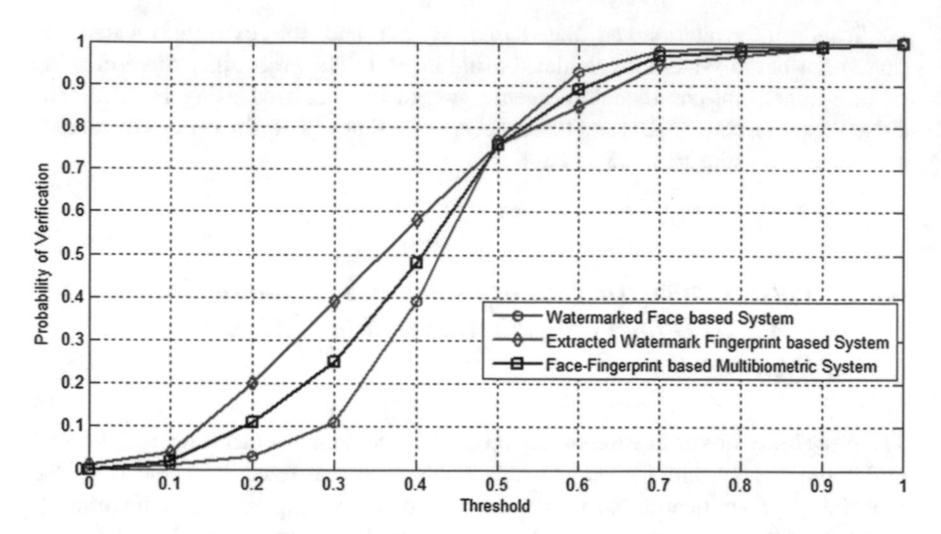

Fig. 6.8 Probability of verification curve for various biometric systems based on proposed technique using DWT-SVD

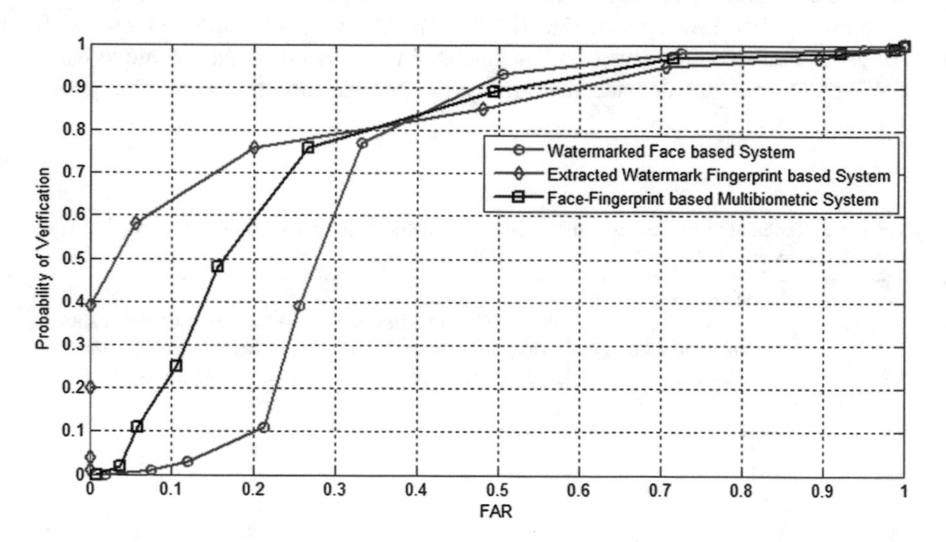

Fig. 6.9 Verification performance curve for various biometric systems based on proposed technique using DWT-SVD

6.4.2 Performance Analysis of Proposed Technique Using DWT-SVD for Authentication Operation of Multibiometric System

The values of False Rejection Rate (FRR) and False Acceptance Rate (FAR) at a different threshold value for watermarked face-based system, extracted watermark fingerprint-based system, and the face-fingerprint-based multibiometric system are

Table 6.4 FRR values and FAR values for biometric systems based on proposed technique using DWT-SVD

Threshold	FRR of FS	FAR of FS	FRR of FPS	FAR of FPS	FRR of MBS	FAR of MBS
0.0	1.000	0.019	1.000	0.000	1.000	0.009
0.1	0.994	0.075	0.950	0.000	0.972	0.038
0.2	0.969	0.119	0.850	0.000	0.909	0.059
0.3	0.881	0.213	0.656	0.000	0.769	0.106
0.4	0.606	0.256	0.494	0.056	0.550	0.156
0.5	0.219	0.331	0.300	0.200	0.259	0.266
0.6	0.075	0.506	0.169	0.481	0.122	0.494
0.7	0.019	0.725	0.075	0.706	0.047	0.716
0.8	0.006	0.950	0.025	0.894	0.016	0.922
0.9	0.006	0.994	0.001	0.981	0.003	0.988
1.0	0.000	1.000	0.000	1.000	0.000	1.000

FS watermarked face-based system, *FPS* extracted watermark fingerprint-based system, *MBS* face-fingerprint-based multibiometric system

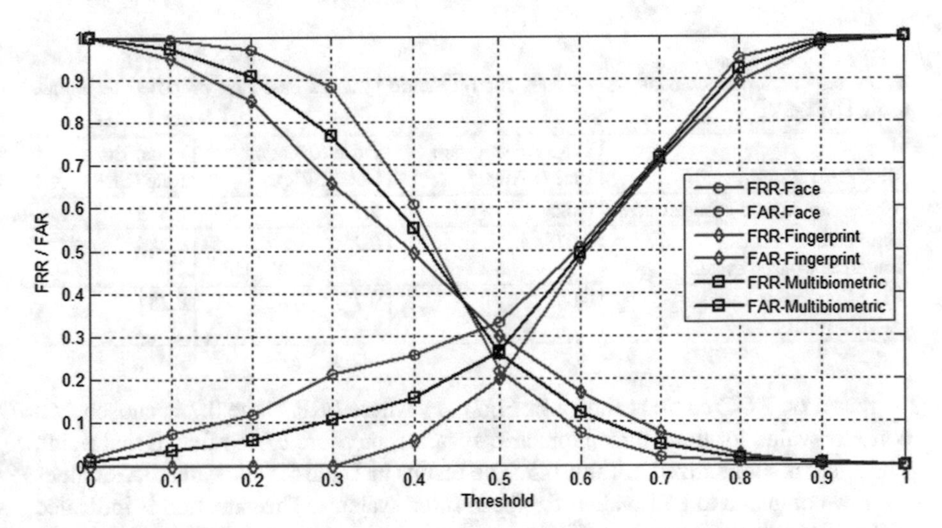

Fig. 6.10 FRR/FAR vs. threshold curve for various biometric systems based on proposed technique using DWT-SVD

summarized in Table 6.4. Based on values in Table 6.4, plot FRR/FAR vs. threshold curve and receiver operating characteristics (ROC) curve for watermarked face-based system, extracted watermark fingerprint-based system, and the face-fingerprint-based multibiometric system is shown in Figs 6.10 and 6.11, respectively. Equal Error Rate (EER) is a point on FRR/FAR vs. threshold curve shown in Fig. 6.11 where FAR and FRR have the same value. The EER value of these three biometric systems is summarized in Table 6.5.

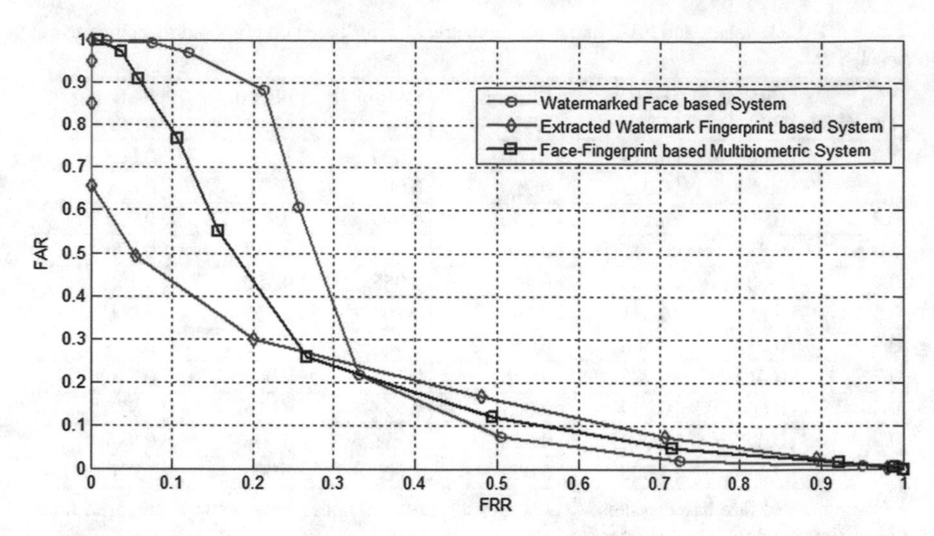

Fig. 6.11 ROC curve for various biometric systems based on proposed technique using DWT-SVD

Table 6.5 Performance evaluation values for biometric systems based on proposed technique using DWT-SVD

Biometric system	False Acceptance Rate (FAR)	False Rejection Rate (FRR)	Equal Error Rate (EER)
Watermarked face-based system	0.025	0.7	0.313
Extracted watermark fingerprint-based system	0.078	0.7	0.268
Face-fingerprint-based multibiometric system	0.052	0.7	0.264

Based on ROC curve is shown in Fig. 6.11 where FRR value 0.7 is chosen as a common value for these three biometric systems, measure FAR value at that point. The result is summarized in Table 6.5. The results in Table 6.5 show that FAR values are low compared to FRR values for these three systems. This situation is indicated that these biometric systems based on proposed technique using DWT-SVD can be used for high security applications.

6.5 Observation of Results

The following are some of the observations made after successfully implementing this proposed watermarking technique for security of biometric image in the multibiometric system:

- This is a hybrid non-blind watermarking technique based on Discrete Wavelet Transform (DWT) and Singular Value Decomposition (SVD).
- This technique is fragile against various watermarking attacks. But this technique is provided robust against flipping watermarking attacks.
- This proposed technique is provided with authentication to the biometric image in the system database of the multibiometric system.
- This technique is provided with security to biometric image against modification attack. It is difficult to generate two biometric images by an impostor, because in this technique, watermark biometric image is encrypted by CS theory, and this encrypted watermark biometric image is embedded into another biometric image.
- This proposed technique does not degrade the verification and authentication performance of the multibiometric system.
- The ROC curve analysis shows that this proposed technique-based multibiometric system can be used in high security applications.
- The limitation of this technique is that wavelet basis matrix, *U* matrix, and *V* matrix of wavelet basis matrix and correct measurement matrix are required for extraction of watermark fingerprint image at extraction side.

6.6 Comparison of Proposed Technique Using DWT-SVD with Existing Techniques Available in Literature

The comparison of the proposed technique with existing techniques available in the literature is summarized in Table 6.6. These techniques are compared using various features and parameters. This proposed technique using DWT-SVD is the embedded sparse data of encrypted biometric data into host biometric image. This proposed watermarking technique is provided with computational security using CS theory procedure, while the existing watermarking techniques are provided with computational security using principal component analysis (PCA), gain factor, and rotation scaling and translation (RST) transform.

Table 6.6 Comparison of proposed watermarking technique using DWT-SVD with existing watermarking techniques available in literature

Features and parameters	Chaudhary technique et al. (2013)	Inamdar technique et al. (2012)	Joshi technique et al. (2011)	Proposed technique
Used SVD coefficients	S values of HH, HL, and LH subband of first level wavelet coefficients	S values of QR decomposition blocks	S values of LL subband of first level wavelet coefficients	S values of HL subband of second level wavelet coefficients
Security achieved	Secret key	PCA	RST transform	CS theory
PSNR (dB)	46.04	44.08	93.29	37.32
SSIM	0.953	0.995	–	0.950

References

Chaudhary, N., Singh, D., & Hussain, D. (2013). Enhancing security of multimodal biometric authentication system by implementing watermarking utilizing DWT and DCT. *IOSR Journal of Computer Engineering, 15*(1), 6–11.

Cox, I., Kilian, J., Shamoon, T., & Leighton, F. (1997). Secure spread spectrum watermarking for multimedia. *IEEE Transactions on Image Processing, 6*(12), 1673–1687.

Ganic, E. and Eskicioglu, A. (2004). *Secure DWT-SVD Domain image watermarking: Embedding data in all frequencies*. ACM Multimedia and Security Workshop 2004, Magdeburg, Germany, pp. 1–9.

Inamdar, V., & Rege, P. (2012). Face features based biometric watermarking of digital image using singular value decomposition for fingerprinting. *International Journal of Security and Its Applications, 6*(2), 47–60.

Jahan, R. (2013). Efficient and secure digital image watermarking scheme using DWT-SVD and optimized genetic algorithm based chaotic encryption. *International Journal of Science, Engineering and Technology Research (IJSETR), 2*(10), 1943–1946.

Jain, A., Prabhakar, S., & Pankanti, S. (1999). *A Filterbank based representation for classification and matching of fingerprint*. International Joint Conference on Neural Networks (IJCNN), Washington, DC, July, pp. 3284–3285.

Joshi, M., Joshi, V., & Raval, M. (2011). Multilevel semi-fragile watermarking technique for improving biometric fingerprint system security. In A. Agrawal, R. C. Tripathi, E. Y.-L. Do, & M. D. Tiwari (Eds.), *Intelligent interactive technologies and multimedia* (pp. 272–283). Berlin/Heidelberg: Springer.

Joshi, V., Raval, M., Rege, P., & Parulkar, S. (2013). Multistage VQ based exact authentication for biometric images. *Computer Society of India (CSI) Journal of Computing, 2*(1–2), R3-25–R3-29.

Kothari, A. (2013). *Design, implementation and performance analysis of digital watermarking for video*. Ph.D. thesis, JJTU, India.

Lu, J., Plataniotis, N., & Venetsanopoulos, A. (2003). Face recognition using LDA based algorithms. *IEEE Transactions on Neural Networks, 14*(1), 195–200.

Mansouri, A., Aznaveh, A., & Azar, F. (2009). SVD based digital image watermarking using complex wavelet transform. *Sadhana © Indian Academy of Science, 34*(3), 393–406.

Prabhakar, S. (2001). *Fingerprint classification and matching using a filterbank*. Ph.D. thesis, Michigan State University, USA.

Shih, F. (2008). *Digital watermarking and steganography – fundamentals and techniques* (pp. 39–41). Boca Raton: CRC Press.

Sreedhanya, A., & Soman, K. (2013). Ensuring security to the compressed sensing data using a steganographic approach. *Bonfring International Journal of Advances in Image Processing, 3* (1), 1–7.

Vidakovic, B. (1999). *Statistical modelling by wavelets* (pp. 115–116). Wiley .

Yan, J. (2009). *Wavelet matrix*. Victoria: Department of Electrical and Computer Engineering, University of Victoria.

Yang, J., Hua, Y., & William, K. (2000). *An efficient LDA algorithm for face recognition*. Proceedings of the International Conference on Automation, Robotics and Computer Vision (ICARCV 2000), pp. 34–47.

Chapter 7
Multibiometric Watermarking Technique Using Fast Discrete Curvelet Transform (FDCuT) and Discrete Cosine Transform (DCT)

Abstract This chapter presents technical details and sparsity property of Curvelet Transform. The hybrid multibiometric watermarking technique using FDCuT-DCT is explained and analyzed in this chapter. The comparison of presented watermarking technique with existing watermarking techniques is also given in this chapter.

7.1 Curvelet Transform

D. Donoho and E. Candes have proposed a sparsity theory based on a new transform for image and signal processing. This new transform is known as a curvelet transform (Candes et al. 2006; Candes and Donoho 2004). The curvelet transform calculates the inner relationship between the image and its curvelet function. This transform is represented by an image into its sparse domain. The curvelet transform is divided into two types: continuous time curvelet transform and discrete time curvelet transform.

7.1.1 Continuous Time Curvelet Transform

The continuous time curvelet transform (Xu et al. 2010; Candes et al. 2006; Candes and Donoho 2004) is expressed by the below equation:

$$c(j, l, k) := \langle f, \varphi_{j,l,k} \rangle \tag{7.1}$$

Here $j = 0, 1, 2\ldots$ is a scale parameter; $l = 0, 1, 2\ldots$ is an orientation parameter; and $k = (k_1, k_2) \in Z^2$ is a translation parameter. The mother curvelet is $\varphi_j(x)$; its Fourier transform is $\varphi_j(\omega) = U_j(\omega)$, where U_j is the frequency window defined in the polar coordinate system:

© Springer International Publishing AG 2018

R. M. Thanki et al., *Multibiometric Watermarking with Compressive Sensing Theory*, Signals and Communication Technology,
https://doi.org/10.1007/978-3-319-73183-4_7

$$U_j(r,\theta) = 2^{-\frac{3j}{4}} W(2^{-j}r) V\left(\frac{2\left[\frac{j}{2}\right]\theta}{2\pi}\right) \tag{7.2}$$

W and *V* are radial and angular windows, respectively, and will always obey certain admissibility conditions. The curvelet at scale 2^{-j}, orientation θ_1, and position $x_k^{(j,l)} = R_{\theta_1}^{-1}\left(k_1 \times 2^{-j}, k_2 \times 2^{-j}\right)$ is expressed by the below equation:

$$\varphi_{j,l,k}(x) = \varphi_j \left\lfloor R_{\theta l}\left(x - x_k^{j,l}\right\rfloor \tag{7.3}$$

So, for $f \in L^2(R^2)$, curvelet transform is expressed as

$$c(i,l,k) := \frac{1}{2\pi^2} \int f\left(\omega\left(\omega_{j,l,k}(\omega \cdot \omega\right) d\omega \tag{7.4}$$

Further, Eq. 7.4 can be expanded as

$$c(i,l,k) := \frac{1}{2\pi^2} \int f\left(\omega\left(\omega_j(R_{\theta l}\omega)\exp\left(i\left\langle x_k^{j,l}, \omega\right\rangle d\omega \tag{7.5}$$

7.1.2 *Discrete Time Curvelet Transform*

The discrete time curvelet transform is linear and takes as input Cartesian arrays of form $f[t_1, t_2]$, $0 \le t_1, t_2 < n$, which allows the output as a collection of coefficients:

$$C^D(j,l,k) := \sum\nolimits_{0 \le t_1, t_2 < n} f[t_1, t_2] \overline{\varphi_{j,l,k}^D[t_1, t_2]} \tag{7.6}$$

where $j = 0, 1, 2\ldots$ is a scale parameter; $l = 0, 1, 2\ldots$ is an orientation parameter; and $k = (k_1, k_2) \in Z^2$ is a translation parameter. The scale parameter depends on the size of the image and is calculated as log2(min $(M, N) - 3$, where *M, N* is the size of row and column of the image. The orientation parameter must be set to a multiplier of 4, and the default value of orientation parameter is 16.

The discrete time curvelet transform has been redesigned with a new mathematical architecture, which is simple and easy to be implemented. It is known as a Fast Discrete Curvelet Transform (FDCuT) (Candes et al. 2006; Candes and Donoho 2004). The Fast Discrete Curvelet Transform is divided into two types: Unequi-Spaced Fast Fourier Transform (USFFT)-based FDCuT and frequency wrapping-based FDCuT.

The USFFT-based FDCuT is an unequal size of the sample, complex, and required for a computational time. While the frequency wrapping-based FDCuT is easy to implement and simpler to understand. The computational time of frequency wrapping-based FDCuT is fast compared to USFFT-based FDCuT. Therefore,

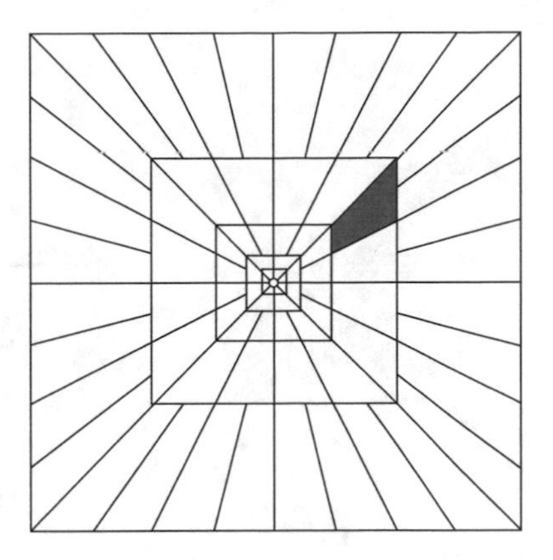

Fig. 7.1 Basic 2D Discrete Curvelet Transform

frequency wrapping-based FDCuT is used by many researchers. The implementing step for frequency wrapping-based FDCuT is given below (Candes et al. 2006; Candes and Donoho 2004; Ying 2005):

- Take FFT of the image.
- Divide FFT into a collection of digital corona tiles (Fig. 7.1).
- For each corona tile
  - Translate the tile to the origin (Fig. 7.2a, b).
  - Wrap the parallelogram-shaped support of the tile around a rectangle centered at the origin (Fig. 7.2c).
  - Take the inverse FFT of the wrapped support.
  - Add the curvelet array to the collection of curvelet coefficients.

When the frequency wrapping-based FDCuT is applied to an image in MATLAB, the image is decomposed into various cells with curvelet coefficients. The size of cells is different and depends on the size of the image. For example, an image with a size of 128 $\times$ 128 pixels is decomposed using frequency wrapping-based FDCuT with 4 scale parameter and 16 orientation parameter into different cells, such as C {1,1} to C {1,4}. The C {1,1} is a low-frequency curvelet coefficient; the other C {1,2} to C {1,4} is high-frequency curvelet coefficient. The size of cells C {1,1} to C {1,3} is less than the actual size of image, but size of C {1,4} is equal to the size of the image. The decomposition of the image using frequency wrapping-based FDCuT with 4 scales and 16 angles is shown in Fig. 7.3a, b. The high-frequency curvelet coefficients C {1,4} are shown in Fig. 7.3c.

- Low-frequency curvelet cell coefficients: $\{1 \times 1\} : C_1^1$
- High-frequency curvelet cell coefficients:

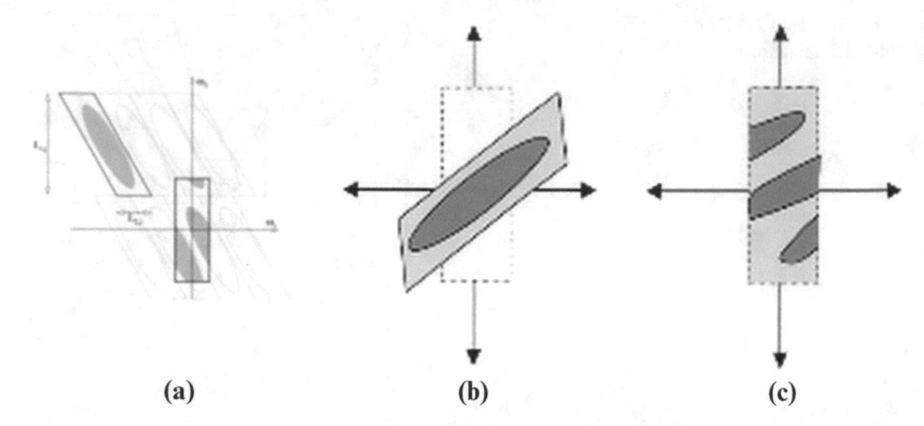

Fig. 7.2 (**a–c**) Wrapping data using FFT to get curvelet coefficients (Candes et al. 2006; Candes and Donoho 2004)

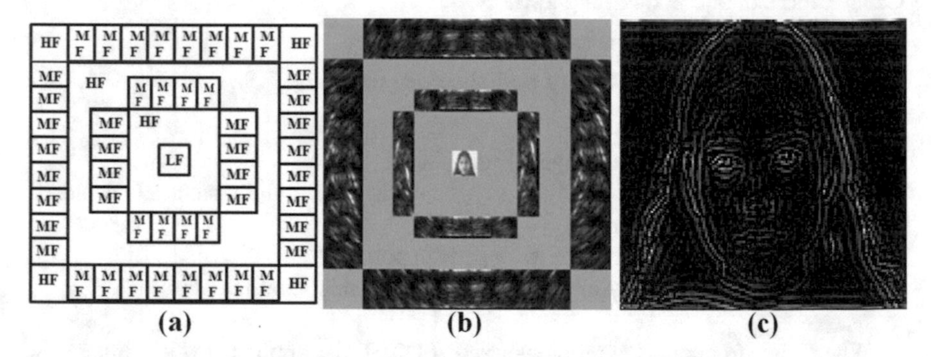

Fig. 7.3 (**a**) Frequency distribution in curvelet coefficients. (**b**) Curvelet coefficients of image. (**c**) High-frequency curvelet coefficients

$$\{1 \times 16\} : C_2^1, C_2^2, C_2^3, C_2^4, C_2^5, \ldots, C_2^{15}, C_2^{16}$$
$$\{1 \times 32\} : C_3^1, C_3^2, C_3^3, C_3^4, C_3^5, \ldots, C_3^{31}, C_3^{32}$$
$$\{1 \times 1\} : C_4^1$$

7.2 Multibiometric Watermarking Technique Using Fast Discrete Curvelet Transform (FDCuT) and Discrete Cosine Transform (DCT)

In this proposed watermarking technique, watermarking is performed in FDCuT domain while encryption of watermark biometric image is performed using DCT basis matrix and CS theory-based encryption process. The DCT basis matrix is used for generation of sparse coefficients of the watermark biometric image. The sparse data of an encrypted watermark biometric image is inserted into high-frequency FDCuT coefficients of a host biometric image. This proposed technique is non-blind technique because the original host biometric image is required at extraction side. This is the reason behind choosing high-frequency curvelet coefficients for the watermarking purpose. Another reason behind choosing high-frequency curvelet coefficients is that the size of these coefficients is equal to the size of sparse data of encrypted watermark biometric image. In this technique, DCT basis matrix is generated using the standard DCT equation (Jain 1999). This DCT basis matrix with its inverse version is multiplying with the watermark biometric image to convert into its sparse coefficients.

This proposed multibiometric watermarking technique is divided into two procedures: watermark biometric encryption and embedding of encrypted watermark biometric and extraction of encrypted watermark biometric and decryption of watermark biometric. The proposed block diagram of multibiometric watermarking technique using FDCuT-DCT is shown in Fig. 7.4.

7.2.1 Watermark Biometric Image Encryption Procedure and Embedding Procedure of Encrypted Watermark Biometric Image

The steps for watermark biometric image encryption procedure and embedding procedure of encrypted watermark biometric image are given below:

- Take a watermark biometric image WBI with a size of $N \times N$, and calculate the size of the image.
- Generate Discrete Cosine Transform (DCT) basis matrix with a size of $N \times N$ using a wavelet basis matrix generation method which is described in Chap. 5.
- The watermark biometric image is converted into its sparse coefficients by multiplying DCT basis matrix with its inverse version with watermark biometric image.

$$x = \Psi_C \times \text{WBI} \times \Psi'_C \tag{7.7}$$

where WBI is a watermark biometric image in terms of the vector, x is sparse coefficients of watermark biometric image, Ψ_C is a DCT basis matrix, and $\Psi_C{}'$ is an inverse DCT basis matrix.

- Generate measurement matrix A with the size of $N \times N$ using a normal distribution with mean $= 0$ and variance $= 1$. This measurement matrix is same for embedder side as well as extraction side.
- The CS measurements of a watermark biometric image are generated by multiplying its sparse coefficients with the measurement matrix.

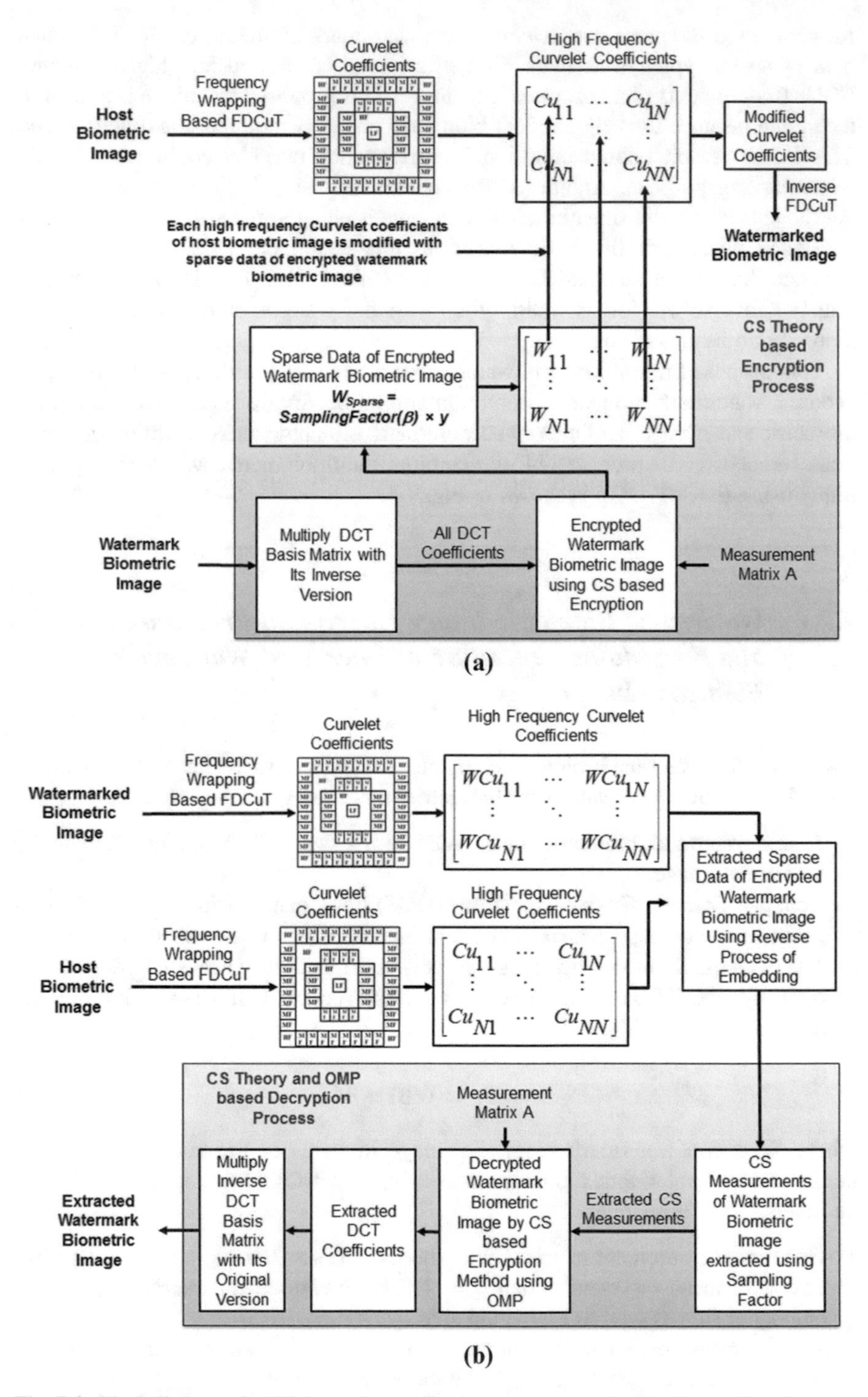

Fig. 7.4 Block diagram of multibiometric watermarking technique using FDCuT-DCT. (**a**) Watermark biometric encryption and embedding of encrypted watermark biometric. (**b**) Extraction of encrypted watermark biometric and decryption of watermark biometric

$$y = A \times x \tag{7.8}$$

where y is an encrypted watermark biometric image in terms of CS measurements and A is a measurement matrix.

- The encrypted watermark biometric image is multiplying with a sampling factor to get sparse data of encrypted watermark biometric image. This sparse data is denoted as W_{Sparse}. This sampling factor is same for embedder side and extraction side.

$$W_{Sparse} = \beta \times y \tag{7.9}$$

where W_{Sparse} is a sparse data of encrypted watermark biometric image, y is an encrypted watermark biometric imager in terms of CS measurements, and β is a sampling factor.

- Take a host biometric image with a size of $N \times N$, and calculate the size of the image.
- The Fast Discrete Curvelet Transform (FDCuT) is applied to the host biometric image to get its curvelet coefficients. The high-frequency curvelet coefficients are choosing for watermark embedding.
- The sparse data of an encrypted watermark image is inserted into high-frequency curvelet coefficients of a host biometric image using multiplicative watermarking equation (Cox et al. 1997; Shih 2008).

$$\text{WCu}_{\text{HFC}} = \text{Cu}_{\text{HFC}} * \left(1 + k \times W_{Sparse}\right) \tag{7.10}$$

where WCu_{HFC} is the modified high-frequency curvelet coefficients of host biometric image, Cu_{HFC} is the original high-frequency curvelet coefficients of host biometric image, W_{Sparse} is the sparse data of encrypted watermark biometric image, and k is a gain factor.

- The inverse frequency wrapping-based FDCuT is applied to modified curvelet high-frequency curvelet coefficients with another unmodified curvelet coefficients of host biometric image to get watermarked biometric image.

7.2.2 Extraction Procedure of Encrypted Watermark Biometric Image and Decryption Procedure for Watermark Biometric Image

The steps for extraction procedure of encrypted watermark biometric image and decryption procedure for watermark biometric image are given below:

- Take a watermarked biometric image which may corrupt or degrade by the impostor. The Fast Discrete Curvelet Transform (FDCuT) is applied to the watermarked biometric image to get its curvelet coefficients.
- Take an original host biometric image. The Fast Discrete Curvelet Transform (FDCuT) is applied to the watermarked biometric image to get its DCT coefficients.
- The sparse data of encrypted watermark biometric image is extracted using the reverse procedure of embedding.

$$W_{Extracted} = \frac{\left(\frac{\mathrm{WCu_{HFC}}}{\mathrm{Cu_{HFC}}} - 1\right)}{k} \tag{7.11}$$

where $\mathrm{WCu_{HFC}}$ is the high-frequency curvelet coefficients of watermarked biometric image, $\mathrm{Cu_{HFC}}$ is the original high-frequency curvelet coefficients of host biometric image, $W_{\mathrm{Extracted}}$ is the extracted sparse data of encrypted watermark biometric image, and k is a gain factor.

- The extracted sparse data of encrypted watermark biometric image is divided by a sampling factor to get extracted encrypted watermark biometric image.

$$y_{Extracted} = \frac{W_{Extracted}}{\beta} \tag{7.12}$$

where $W_{\mathrm{Extracted}}$ is the extracted sparse data of encrypted watermark biometric image, $y_{\mathrm{Extracted}}$ is the extracted encrypted watermark biometric image, and β is a sampling factor.

- After extracting the encrypted watermark biometric image, the decryption of watermark biometric image process is performed using CS theory and orthogonal matching pursuit (OMP) algorithm.
- The OMP is applied to the extracted encrypted watermark biometric image with correct measurement matrix to get sparse coefficients. These coefficients are all DCT coefficients of a watermark biometric image.

$$x' = \mathrm{OMP}(y_{Extracted}, A) \tag{7.13}$$

where x' is an extracted all DCT coefficient of watermark biometric image.

- Finally, the inverse DCT basis matrix with its original version is multiplying with extracted wavelet coefficients of a watermark biometric image to get extracted watermark biometric image at extraction side.

$$\mathrm{EWBI} = \Psi'_C \times x' \times \Psi_C \tag{7.14}$$

where EWBI is an extracted watermark biometric image.

- After extracting watermark biometric image, two hypotheses are formulated for authentication of host biometric image.
  - If $SSIM(\text{WBI}, EWBI) > \tau$, then host biometric image is authenticated.
  - If $SSIM(\text{WBI}, EWBI) < \tau$, then host biometric image is unauthenticated.

where τ is a predefined threshold value for a decision about an authenticity of host biometric image.

7.3 Experimental Results

For testing and analysis of this proposed technique, 8-bit grayscale 50 face images from Indian face database and 110 face images from FEI face database are taken as host biometric images. The 8-bit grayscale 80 fingerprint images from FVC2002 DB3 setB and 80 fingerprint images from FVC2004 DB4 setB are taken as watermark biometric images. The size of host face image and watermark fingerprint image is 128 × 128 pixels. The sparse data of encrypted watermark fingerprint image is inserted into host face image using the below procedure.

First, Discrete Cosine Transform (DCT) basis matrix is generated with a size of 128 × 128 using steps mentioned in Eq. 5.3 in Chap. 5. The DCT basis matrix with its inverse version is multiplying with the watermark fingerprint image to get sparse coefficients x of a watermark fingerprint image with a size of 128 × 128. The measurement matrix A with the size of 128 × 128 is generated using a Gaussian distribution with mean = 0 and variance = 1. The CS measurements of watermark fingerprint image with a size of 128 × 128 are generated using $y_{128 \times 128} = A_{128 \times 128} \times x_{128 \times 128}$. The encrypted watermark fingerprint image is multiplying with sampling factor to get sparse data of encrypted watermark fingerprint image and denoted as W_{Sparse}. The sparse data W_{Sparse} of encrypted watermark fingerprint image is inserted into high-frequency curvelet coefficients of host face image using gain factor at embedder side. On the extraction side, sparse data $W_{\text{Extracted}}$ of encrypted watermark fingerprint image is extracted using reverse procedure of embedding. The decryption of watermark fingerprint image from extracted encrypted fingerprint image can be done by using the below procedure at extraction side.

For decryption of watermark fingerprint image, the input of OMP algorithm is encrypted watermark fingerprint image with a size of 128×128 and measurement matrix A with a size of 128×128. The output of OMP algorithm is sparse coefficients of watermark fingerprint image with a size of 128×128. The inverse DCT basis matrix with its original version is multiplying with these extracted sparse coefficients to get extracted watermark fingerprint image at extraction side. Figure 7.5 shows original face image and watermarked face image, original fingerprint image and extracted fingerprint image, and encrypted fingerprint image and extracted

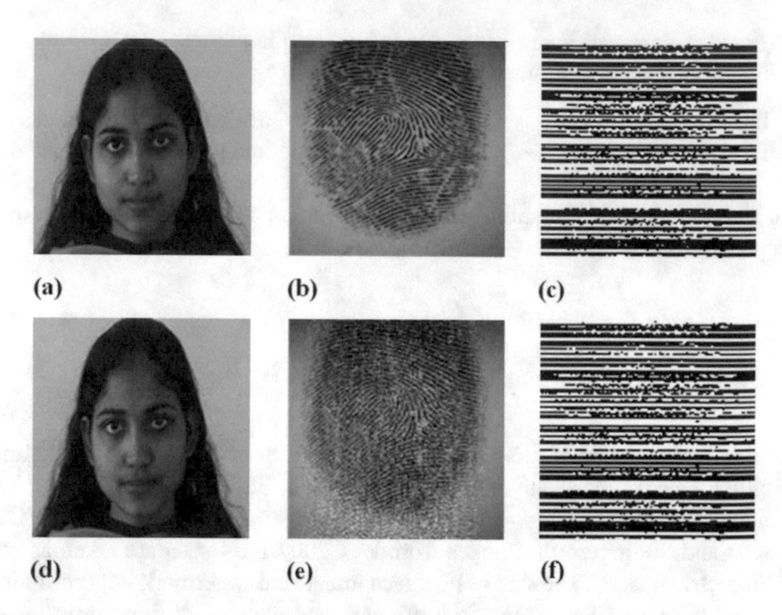

Fig. 7.5 Experimental results of multibiometric watermarking technique using FDCuT-DCT for gain factor $k = 0.2$. (**a**) Host face image. (**b**) Watermark fingerprint image. (**c**) Encrypted watermark fingerprint image. (**d**) Watermarked face image. (**e**) Extracted watermark fingerprint image. (**f**) Extracted encrypted watermark fingerprint image

fingerprint image. These results are generated using gain factor k value 0.2 and sampling factor β value 0.0001.

For testing of the nature of proposed watermarking technique, various watermarking attacks such as JPEG compression with different quality factor ($Q = 90$–50), the addition of noise (Gaussian noise, speckle noise, salt and pepper), applied different image filter with different filter mask size (median filter, mean filter, and Gaussian low-pass filter), sharpening, blurring, histogram equalization, and different geometric attacks like flipping, rotation, and cropping are applied on watermarked face image. The performance result of watermarking attacks on proposed watermarking technique is shown in Fig. 7.6.

The quality measure such as peak signal to noise ratio (PSNR) is used for quality check between original face and watermarked face image. The structural similarity index measure (SSIM) is used for quality check between the original watermark fingerprint and extracted watermark fingerprint image in proposed multibiometric watermarking technique. The results of multibiometric watermarking technique using FDCuT-DCT under various watermarking attacks are summarized in Table 7.1.

For authentication of the host face image of an individual, SSIM value between the watermark fingerprint image and extracted watermark fingerprint image must be greater than the predefined threshold value $\tau = 0.90$. SSIM values in Table 7.1 are indicated that when a watermarking attack is applied on watermarked face image, then the watermark fingerprint image is not extracted successfully at extraction side

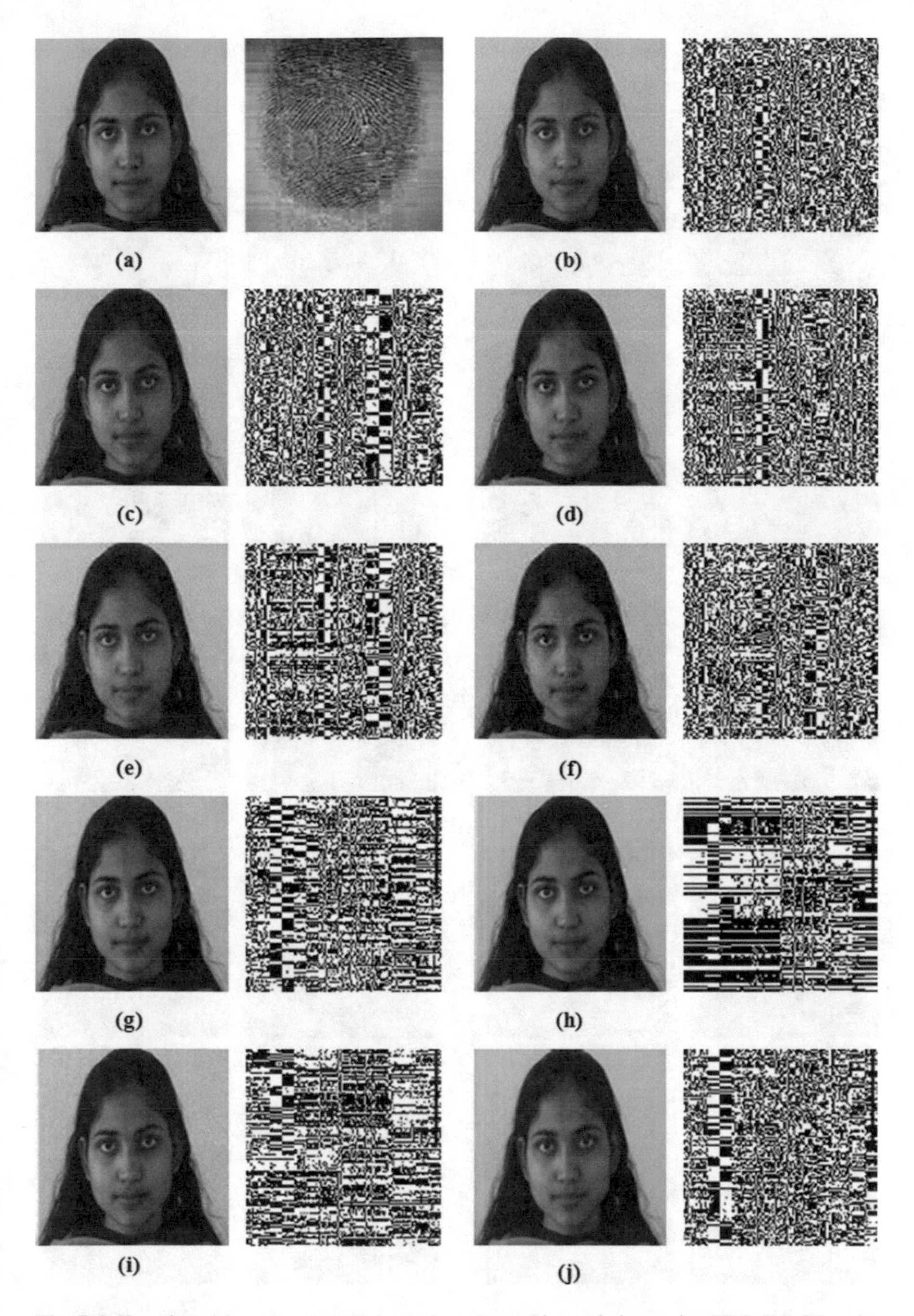

Fig. 7.6 Experimental results of multibiometric watermarking technique using FDCuT-DCT under various watermarking attacks for gain factor $k = 0.2$. (**a**) No attack. (b) JPEG compression (Q = 90). (**c**) JPEG compression (Q = 80). (**d**) JPEG compression (Q = 70). (**e**) JPEG compression (Q = 60). (**f**) JPEG compression (Q = 50). (**g**) Gaussian noise attack. (**h**) Salt and pepper noise attack. (**i**) Speckle noise attack. (**j**) Median filter attack (3 × 3). (**k**) Median filter attack (5 × 5). (**l**) Median filter attack (7 × 7). (**m**) Mean filter attack (3 × 3). (**n**) Mean filter attack (5 × 5). (**o**) Mean filter attack (7 × 7). (**p**) Gaussian low-pass filter attack (3 × 3). (**q**) Gaussian low-pass filter attack (5 × 5). (**r**) Gaussian low-pass filter attack (7 × 7). (**s**) Sharpening attack. (**t**) Blurring attack. (**u**) Histogram equalization attack. (**v**) Flipping attack. (**w**) Rotation attack (90°). (**x**) Cropping attack (20 %)

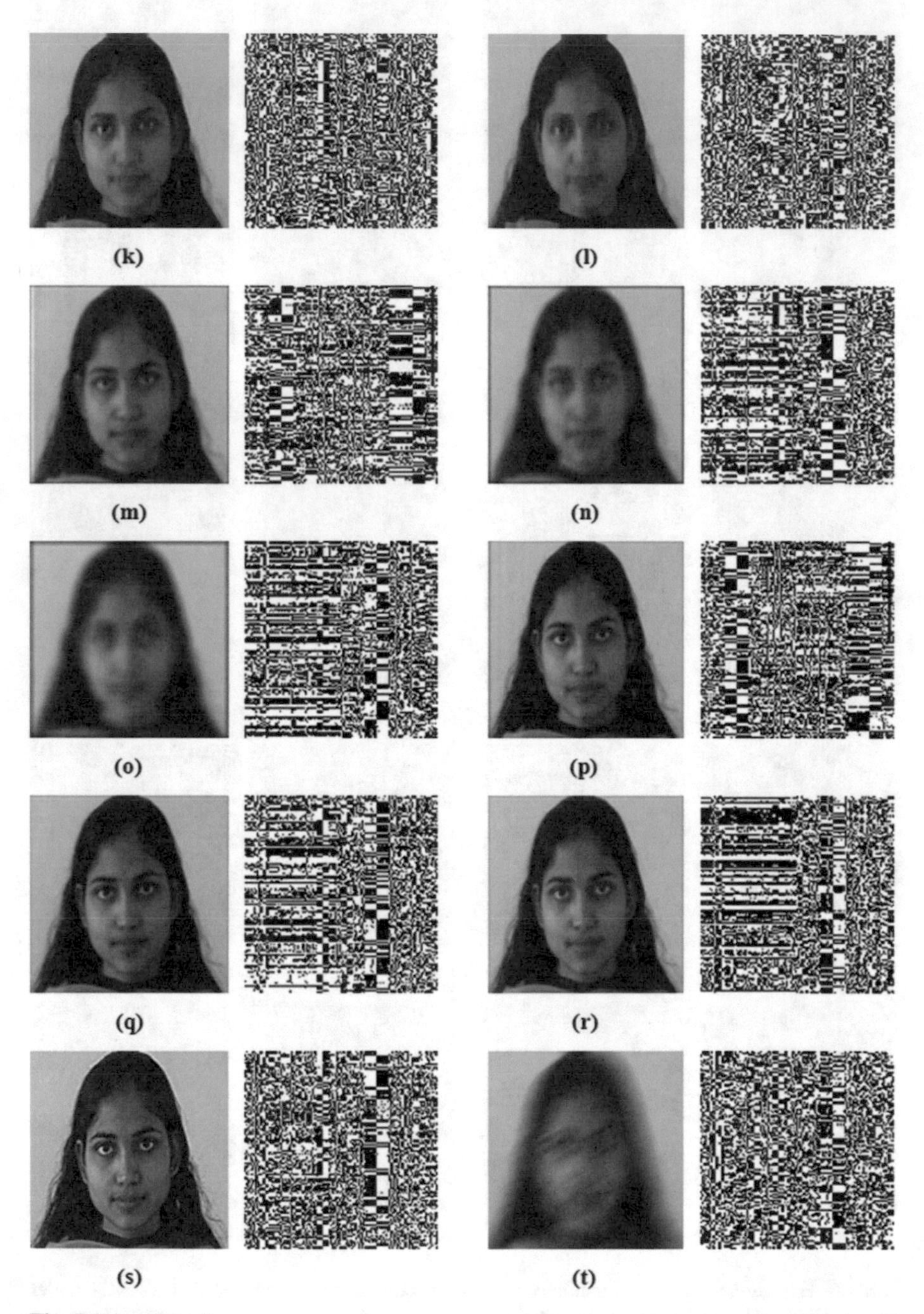

Fig. 7.6 (continued)

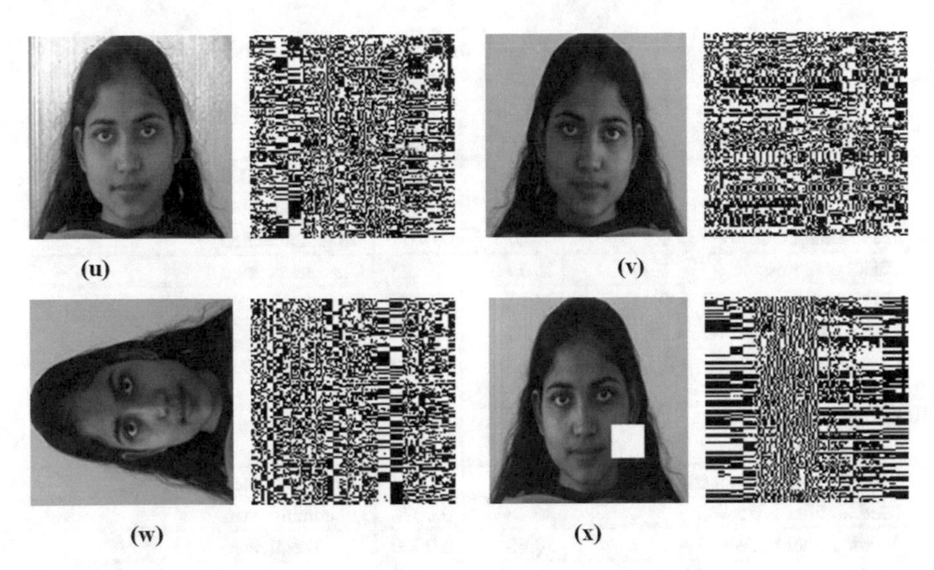

Fig. 7.6 (continued)

and SSIM value is less than for 0.90 for all watermarking attacks. This situation indicated that this proposed multibiometric watermarking technique using FDCuT-DCT is fragile in nature against watermarking attacks.

In the watermark embedding process, the gain factor is multiplying with sparse data of encrypted watermark fingerprint image, and the resultant values are inserted into the wavelet coefficients of host face image. For extraction of encrypted watermark fingerprint image from watermarked face image, gain factor is also required at extraction. Thus, gain factor has significant on PSNR value of watermarked face image as well as SSIM value of extracted watermark fingerprint image. Table 7.2 shows the effect of the gain factor on PSNR value of the watermarked face image and SSIM value of the extracted watermark fingerprint image.

The gain factor value should be 0.1–10 for this proposed multibiometric watermarking technique using FDCuT-DCT. Because when the gain factor is set to 11 or more, then the PSNR value is less than 28 dB. This situation indicates that gain factor does not set above 11 for this proposed technique. The value of PSNR, value below 28 dB, is not acceptable for human visual system (HVS) property of watermarking technique.

The reason behind achieving fragility for this proposed technique is that the encrypted watermark fingerprint image is destroyed when a watermarking attack is applied on watermarked face image. The encrypted watermark fingerprint image is destroyed because high-frequency curvelet coefficients are considered for embedding it. Figure 7.7 shows the histogram of encrypted watermark fingerprint image and extracted encrypted watermark fingerprint image without watermarking attacks and under watermarking attacks. Figure 7.7a indicated the histogram of encrypted watermark fingerprint image, and extracted encrypted watermark fingerprint is the

Table 7.1 PSNR (dB) values and SSIM values for multibiometric watermarking technique using FDCuT-DCT under various watermarking attacks for gain factor $k = 0.2$

Attack	PSNR (dB)	SSIM	Decision about authentication of host face image
No attack	64.07	0.9861	Authenticated
JPEG compression ($Q = 90$)	39.54	0.0009	Unauthenticated
JPEG compression ($Q = 80$)	36.67	0.0041	Unauthenticated
JPEG compression ($Q = 70$)	35.17	0.0123	Unauthenticated
JPEG compression ($Q = 60$)	34.05	0.0024	Unauthenticated
JPEG compression ($Q = 50$)	33.22	0.0057	Unauthenticated
Gaussian noise (mean = 0, Variance = 0.0001)	37.63	0.0019	Unauthenticated
Salt and pepper noise (variance = 0.0005)	36.61	0.0025	Unauthenticated
Speckle noise (variance = 0.0004)	37.50	0.0017	Unauthenticated
Median filter (size = 3 × 3)	36.28	0.0076	Unauthenticated
Median filter (size = 5 × 5)	30.83	0.0000	Unauthenticated
Median filter (size = 7 × 7)	28.14	0.0014	Unauthenticated
Mean filter (size = 3 × 3)	25.04	0.0078	Unauthenticated
Mean filter (size = 5 × 5)	21.99	0.0008	Unauthenticated
Mean filter (size = 7 × 7)	20.18	0.0009	Unauthenticated
Gaussian low-pass filter (size = 3 × 3)	34.11	0.0038	Unauthenticated
Gaussian low-pass filter (size = 5 × 5)	34.12	0.0098	Unauthenticated
Gaussian low-pass filter (size = 7 × 7)	34.09	0.0017	Unauthenticated
Sharpening	30.52	0.0008	Unauthenticated
Blurring	19.64	0.0011	Unauthenticated
Histogram equalization	19.47	0.0026	Unauthenticated
Flipping	13.89	0.0010	Unauthenticated
Rotation (90°)	6.91	0.0041	Unauthenticated
Cropping (20%)	16.17	0.0016	Unauthenticated

Table 7.2 Effect of gain factor k on performance of multibiometric watermarking technique using FDCuT-DCT

Gain factor	PSNR (dB)	SSIM
0.1	69.37	0.9861
0.2	64.07	0.9861
0.5	52.23	0.9870
0.7	51.63	0.9869
1	47.90	0.9857
2	43.56	0.9863
5	35.25	0.9864
7	31.99	0.9860
10	28.66	0.9868
12	25.61	0.9864

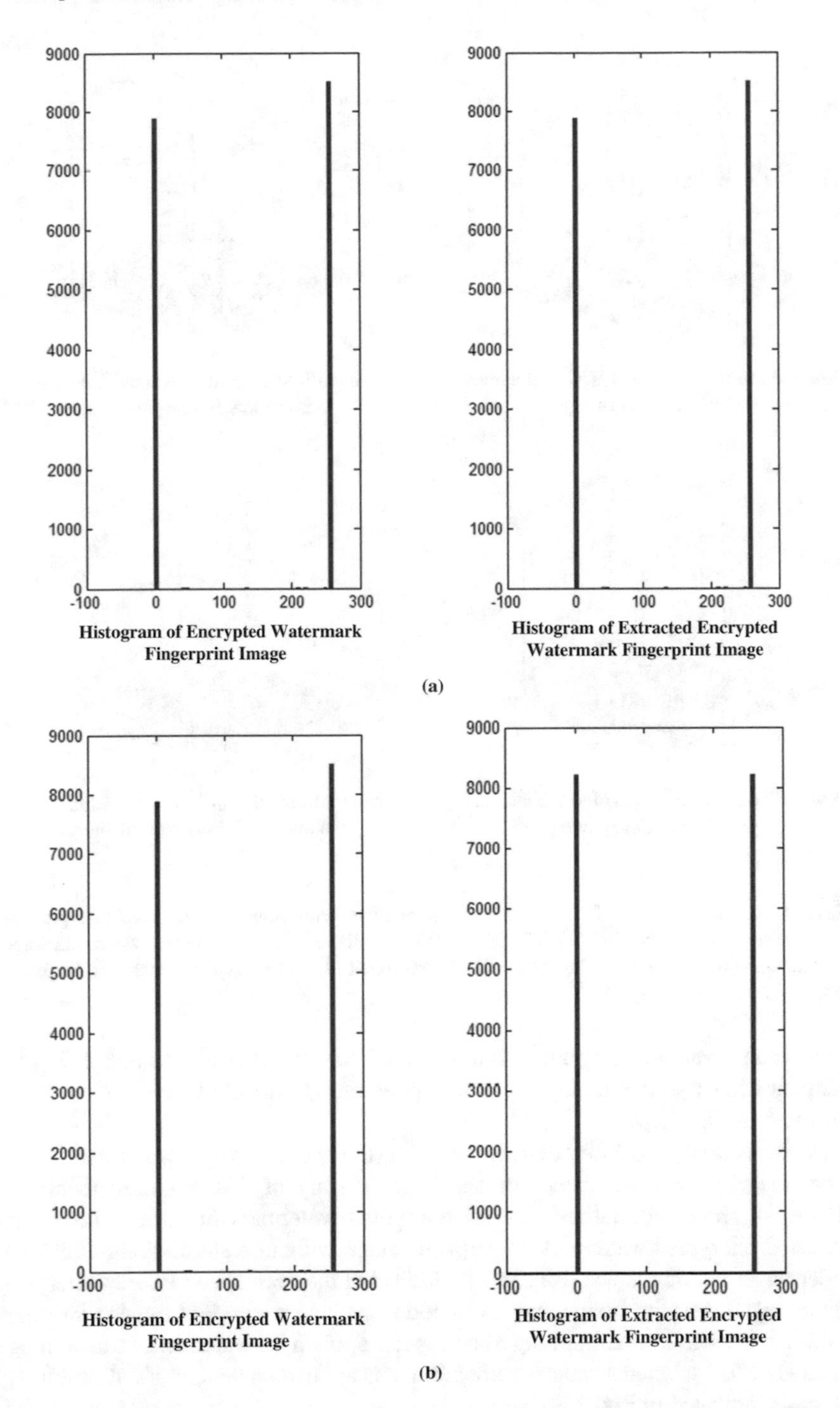

Fig. 7.7 Histogram of encrypted watermark fingerprint image in multibiometric proposed watermarking technique using FDCuT-DCT. (**a**) Histogram of encrypted watermark fingerprint image without watermarking attack. (**b**) Histogram of encrypted watermark fingerprint image with watermarking attack

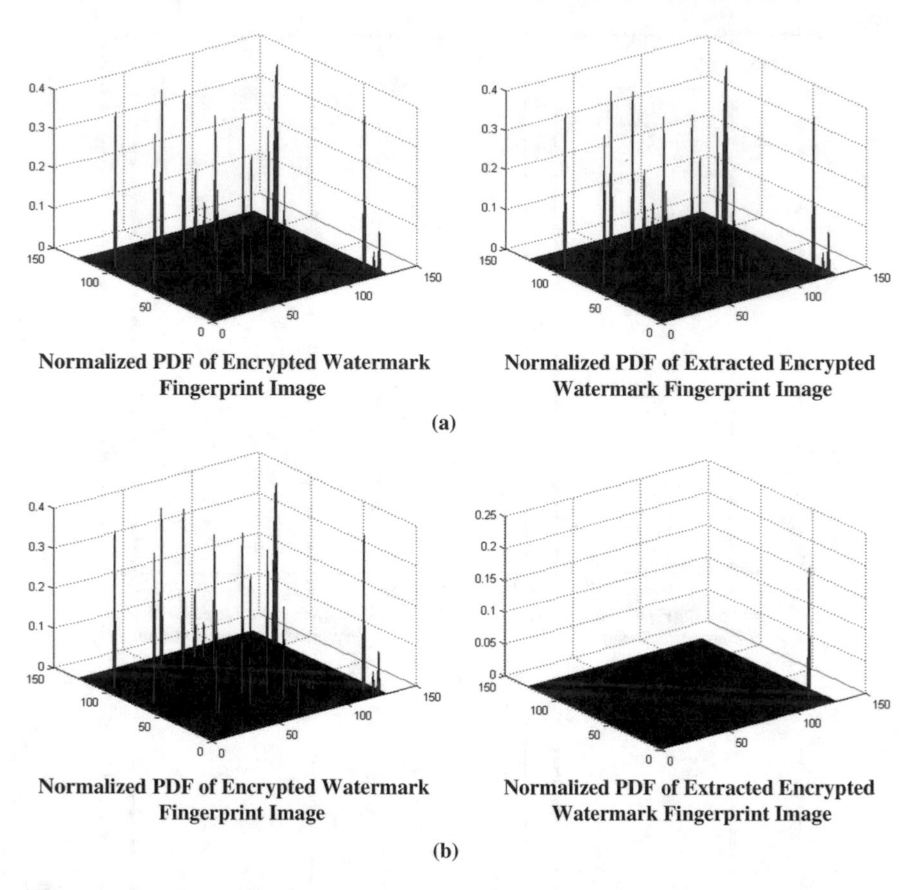

Fig. 7.8 Normalized PDF of encrypted watermark fingerprint image in multibiometric proposed watermarking technique using FDCuT-DCT. (**a**) Normalized PDF of encrypted watermark fingerprint image without watermarking attack. (**b**) Normalized PDF of encrypted watermark fingerprint image with watermarking attack

same when a watermarking attack is not applied on watermarked face image. But it is different when a watermarking attack is applied on watermarked face image which is indicated in Fig. 7.7b.

A normalized probability density function (PDF) of encrypted watermark fingerprint image is also calculated for checking fragility of this proposed technique. Figure 7.8 shows normalized PDF of encrypted watermark fingerprint image and extracted encrypted watermark fingerprint image without watermarking attack and under watermarking attack. Figure 7.8a indicated the normalized PDF of encrypted watermark fingerprint image, and extracted encrypted watermark fingerprint image is the same when a watermarking attack is not applied on watermarked face image. But it is different when a watermarking attack is applied on watermarked face image which is indicated in Fig. 7.8b.

7.4 Analysis of Effect of Proposed Technique Using FDCuT-DCT on Performance of Multibiometric System

In any biometric system, two operations such as verification and authentication of an individual are very important. The performance of any biometric system is measured based on these two operations so that any biometric image protection technique should not degrade the performance of these two operations of biometric system. Therefore, in this chapter, firstly, performance analysis of individual watermarked face image-based system and individual extracted watermark fingerprint image-based system is performed. Finally, performance analysis of face-fingerprint-based multibiometric system is performed. This multibiometric system is made by watermarked face-based system and extracted watermark fingerprint-based system.

In this technique, fingerprint information is embedded into the face image. Therefore, in this section, it is checked that insertion of fingerprint information should not change the performance of face-based biometric system. In this section, it is also checked that performance of fingerprint-based biometric system should not change due to CS theory-based encryption process and decryption process.

In order to showcase the effect of watermark fingerprint image on the host face image, face recognition algorithm developed by various researchers (Yang et al. 2000; Lu et al. 2003) is used. In order to showcase the effect of CS theory-based encryption and decryption on watermark fingerprint image, fingerprint recognition algorithm developed by various researchers (Jain et al. 1999; Prabhakar 2001) is used. These two algorithms are selected because the output of the algorithms is given the Euclidean distance between query biometric image and its closest match in the database.

For analysis of verification and authentication performance of the multibiometric system, first, 160 watermarked face images and 160 extracted fingerprint images are stored in the system database of watermarked face-based system and extracted watermark fingerprint-based system, respectively. These watermarked face images and extracted fingerprint images are generated using 50 images from Indian face database, 110 face images from FEI face database, 80 images from FVC2002 DB3 setB, and 80 images from FVC2004 DB4 setB.

For testing of multibiometric system using this proposed technique, 50 images from Indian face database and 110 face images from FEI face database are taken as authentic face images. Also, 50 images from FEI face database and 110 face images from CVL face database are taken as fake face images. These face images are taken as query images for watermarked face-based system. The 80 images from FVC2002 DB3 setB and 80 images from FVC2004 DB4 setB are taken as authentic fingerprint images. Also, 80 images from FVC2002 DB4 setB and 80 images from FVC2004 DB3 setB are taken as fake fingerprint images. These fingerprint images are taken as query images for the extracted watermark fingerprint-based system.

Based on the matching score obtained by recognition algorithms (Yang et al. 2000; Lu et al. 2003; Jain et al. 1999; Prabhakar 2001), the average Euclidean

Table 7.3 Average Euclidean distance for biometric systems based on proposed technique using FDCuT-DCT (for 160 Images)

Average Euclidean distance for watermarked face-based system	Between genuine database and watermarked database	18.06
	Between fake database and watermarked database	511.63
Average Euclidean distance for extracted watermark fingerprint-based system	Between genuine database and extracted database	618.80
	Between fake database and extracted database	779.19
Average Euclidean distance for face-fingerprint based multibiometric system	Between genuine database and watermarked-extracted database	349.93
	Between fake database and watermarked-extracted database	645.51

distance for individual watermarked face-based system, individual watermark fingerprint-based system, and face-fingerprint-based multibiometric system are calculated. The average results for these systems are summarized in Table 7.3.

The threshold distance selected based on this is 450. The average threshold value between a fake biometric database with watermarked face biometric database and extracted fingerprint biometric database is calculated. The average distance value of the impostor biometric database is 645.51. This distance is greater than the selected threshold distance. The average distance value between a genuine face database and genuine fingerprint database with watermarked face biometric database and extracted fingerprint biometric database is also calculated. The average distance value of the genuine biometric database is 349.93. This distance is less than the selected threshold distance. Since the average distance between genuine multibiometric database and its watermarked and extracted database is less than selected threshold distance, this situation indicates that the performance of biometric database of multibiometric system remains unaffected due to this proposed technique.

The effect of this proposed technique on verification operation and authentication operation of the multibiometric system can be analyzed by various parameters. These parameters such as the probability of verification, False Rejection Rate (FRR), False Acceptance Rate (FAR), and Equal Error Rate (EER) are used for evaluation of the performance of the multibiometric system. The probability of verification of watermarked face-based system and the extracted watermark fingerprint-based system is calculated using Eq. 1.1. The probability of verification of face-fingerprint-based multibiometric system is calculated using Eq. 1.2. The False Rejection Rate (FRR) and False Acceptance Rate (FAR) for various thresholds are calculated using Eq. 1.3 and Eq. 1.4.

7.4.1 *Performance Analysis of Proposed Technique Using FDCuT-DCT for Verification Operation of Multibiometric System*

The probability of verification of this proposed technique for face-fingerprint-based multibiometric system for various thresholds is summarized in Table 7.4. The probability of verification curve for this proposed technique for face-fingerprint-based multibiometric system for various thresholds is shown in Fig. 7.9. The

Table 7.4 Probability of verification values for biometric systems based on proposed technique using FDCuT-DCT

Threshold	Probability of verification of watermarked face-based system	Probability of verification of extracted watermark fingerprint-based system	Probability of verification of face-fingerprint-based multibiometric system
0.0	0.000	0.000	0.000
0.1	0.000	0.050	0.025
0.2	0.056	0.150	0.103
0.3	0.569	0.344	0.456
0.4	0.831	0.506	0.669
0.5	0.888	0.700	0.794
0.6	0.938	0.831	0.884
0.7	0.975	0.925	0.950
0.8	0.975	0.975	0.975
0.9	0.988	0.994	0.991
1.0	1.000	1.000	1.000

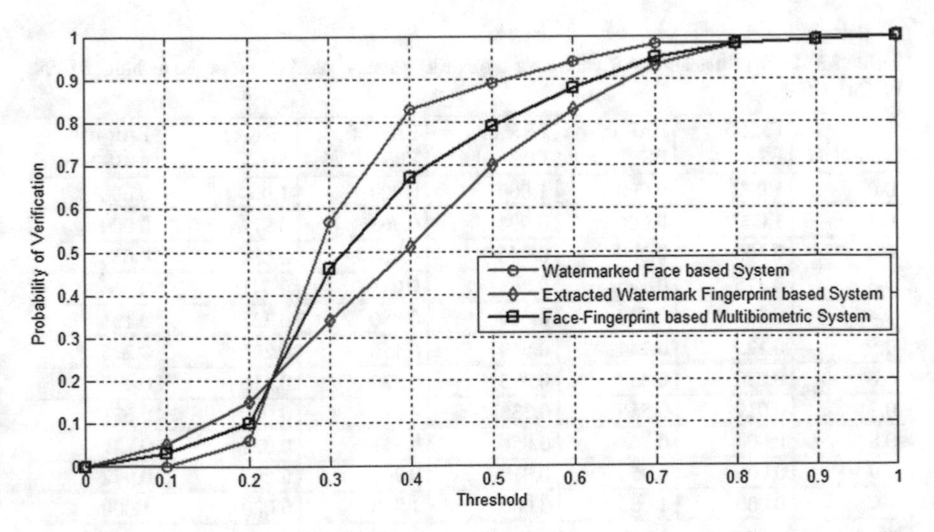

Fig. 7.9 Probability of verification curve for various biometric systems based on proposed technique using FDCuT-DCT

verification performance curve for this proposed technique for face-fingerprint-based multibiometric system is shown in Fig. 7.10. It plots the probability of verification versus False Acceptance Rate (FAR). The FAR values for different thresholds are given in Table 7.5. This curve indicates that this proposed technique using FDCuT-DCT does not degrade the verification performance of the multibiometric system.

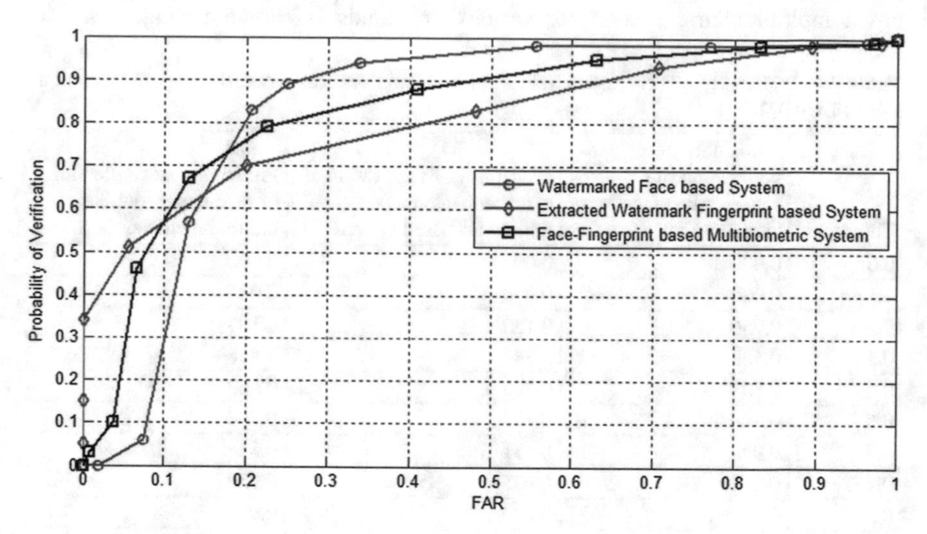

Fig. 7.10 Verification performance curve for various biometric systems based on proposed technique using FDCuT-DCT

Table 7.5 FRR values and FAR values for biometric systems based on proposed technique using FDCuT-DCT

Threshold	FRR of FS	FAR of FS	FRR of FPS	FAR of FPS	FRR of MBS	FAR of MBS
0.0	1.000	0.000	1.000	0.000	1.000	0.000
0.1	1.000	0.006	0.994	0.000	0.972	0.009
0.2	0.606	0.069	0.956	0.000	0.884	0.038
0.3	0.375	0.125	0.863	0.031	0.541	0.066
0.4	0.225	0.200	0.669	0.056	0.325	0.131
0.5	0.125	0.244	0.519	0.156	0.206	0.225
0.6	0.050	0.338	0.375	0.325	0.116	0.409
0.7	0.019	0.556	0.238	0.613	0.050	0.631
0.8	0.006	0.763	0.056	0.831	0.025	0.831
0.9	0.006	0.956	0.019	0.963	0.007	0.972
1.0	0.000	1.000	0.000	1.000	0.000	1.000

FS watermarked face-based system, *FPS* extracted watermark fingerprint-based system, *MBS* face-fingerprint-based multibiometric system

7.4.2 Performance Analysis of Proposed Technique Using FDCuT-DCT for Authentication Operation of Multibiometric System

The values of False Rejection Rate (FRR) and False Acceptance Rate (FAR) at a different threshold value for watermarked face-based system, extracted watermark fingerprint-based system, and the face-fingerprint-based multibiometric system is summarized in Table 7.5. Based on values in Table 7.5, plot FRR/FAR vs. threshold curve and receiver operating characteristics (ROC) curve for watermarked face-based system, extracted watermark fingerprint-based system, and the face-fingerprint-based multibiometric system is shown in Figs. 7.11 and 7.12, respectively. Equal Error Rate (EER) is a point on FRR/FAR vs. threshold curve shown in Fig. 7.12 where FAR and FRR have the same value. The EER value of these three biometric systems is summarized in Table 7.6.

Based on ROC curve which is shown in Fig. 7.12 where FRR value 0.7 is chosen as a common value for these three biometric systems, measure FAR value at that point. The result is summarized in Table 7.6. The results in Table 7.6 show that FAR values are low compared to FRR value for these three systems. This situation indicates that these biometric systems based on proposed technique using FDCuT-DCT can be used for high security applications.

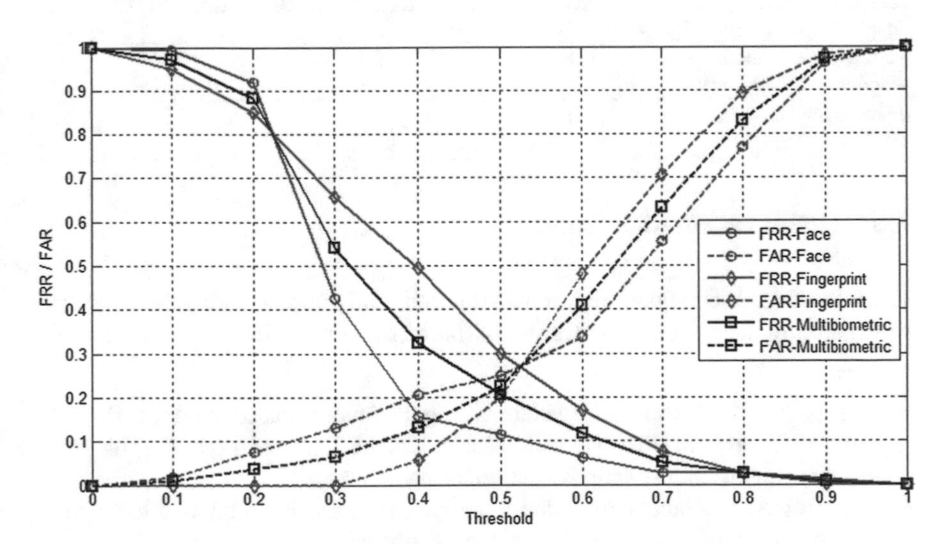

Fig. 7.11 FRR/FAR vs. threshold curve for various biometric systems based on proposed technique using FDCuT-DCT

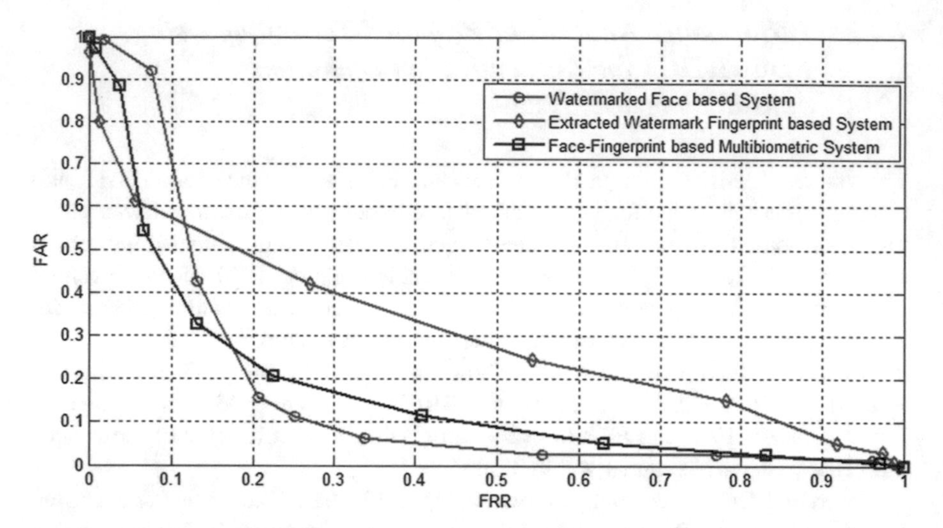

Fig. 7.12 ROC curve for various biometric systems based on proposed technique using FDCuT-DCT

Table 7.6 Performance evaluation values for biometric systems based on proposed technique using FDCuT-DCT

Biometric system	False Acceptance Rate (FAR)	False Rejection Rate (FRR)	Equal Error Rate (EER)
Watermarked face-based system	0.025	0.7	0.195
Extracted watermark fingerprint-based system	0.182	0.7	0.361
Face-fingerprint-based multibiometric system	0.041	0.7	0.217

7.5 Observation of Results

The following observations are made after successfully implementing this proposed watermarking technique for security of biometric image in the multibiometric system.

- This is a hybrid non-blind watermarking technique based on Fast Discrete Curvelet Transform (FDCuT) and Discrete Cosine Transform (DCT). This technique is fragile against various watermarking attacks.
- This proposed technique provided more payload capacity compared to existing watermarking techniques available in the literature.
- This proposed technique provided authentication to the biometric image in the system database of the multibiometric system.
- This proposed technique provided security to biometric image against modification attack. It is difficult to generate two biometric images by an impostor. Because in this technique, watermark biometric image is encrypted by CS theory,

and this encrypted watermarked biometric image is embedded into host biometric image.
- This proposed technique does not degrade the verification and authentication performance of the multibiometric system.
- The ROC curve analysis shows that this proposed technique-based multibiometric system can use in applications where high security is required.
- The limitation of this technique is that DCT basis matrix and correct measurement matrix are required for extraction of watermark fingerprint image at extraction side.

7.6 Comparison of Proposed Technique Using FDCuT-DCT with Existing Techniques Available in Literature

The comparison of proposed technique with existing techniques available in the literature is summarized in Table 7.7. These techniques are compared using various features and parameters. This proposed technique using FDCuT-DCT embedded sparse data of encrypted biometric data into host biometric image. This proposed watermarking technique is fragile while existing watermarking techniques in literature are robust. This proposed watermarking technique provided computational security using CS theory procedure. The existing watermarking techniques provided computational security using gain factor and Arnold scrambling algorithm.

The PSNR value in the Bazargani technique (Bazargani et al. 2012) is 60.80 dB, in Xu technique (Xu et al. 2010) 39.85 dB, in Zhang technique (Zhang et al. 2008) 43.18 dB, and in the proposed technique 64.07 dB. The SSIM value in the Bazargani technique (Bazargani et al. 2012) is 0.971, in Xu technique (Xu et al. 2010) 0.962,

Table 7.7 Comparison of proposed watermarking technique using FDCuT-DCT with existing watermarking techniques available in literature

Features and parameters	Bazargani technique et al. (2012)	Xu technique et al. (2010)	Zhang technique et al. (2008)	Proposed technique
Type of technique	Robust	Robust	Robust	Fragile
Used host medium	Standard image	Fingerprint image	Standard image	Face image
Used watermark	Logo	Random numbers	Face image	Sparse measurements of fingerprint image
Used curvelet coefficients	Low frequency	Low frequency	Low frequency	High frequency
Security achieved	No such scope	Arnold transform	No such scope	CS theory
PSNR (dB)	60.80	39.85	43.18	64.07
SSIM	0.971	0.962	–	0.986

and in the proposed technique 0.986. These results indicate that performance of the proposed technique is better than existing techniques in terms of imperceptibility, robustness, and security.

References

Bazargani, M., Ebrahimi, H., & Dianat, R. (2012). Digital image watermarking in wavelet, contourlet and curvelet domains. *Journal of Basic and Applied Scientific Research, 2*(11), 11296–11308.

Candes, E., & Donoho, D. (2004). New tight frames of curvelets and optimal representations of objects with piecewise-C2 singularities. *Communications On Pure and Applied Mathematics, 57*, 219–266.

Candes, E., Demanet, L., Donoho, D., & Ying, L. (2006). Fast discrete curvelet transforms. *SIAM Multiscale Modeling & Simulation, 5*(3), 861–889.

Cox, I., Kilian, J., Shamoon, T., & Leighton, F. (1997). Secure spread spectrum watermarking for multimedia. *IEEE Transactions on Image Processing, 6*(12), 1673–1687.

Jain, A. (1999). *Fundamentals of digital image processing*. Upper Saddle River: Prentice Hall Inc..

Jain, A., Prabhakar, S., & Pankanti, S. (1999). *A Filterbank based representation for classification and matching of fingerprint.* International Joint Conference on Neural Networks (IJCNN), Washington, DC, July, pp. 3284–3285.

Lu, J., Plataniotis, N., & Venetsanopoulos, A. (2003). Face recognition using LDA based algorithms. *IEEE Transactions on Neural Networks, 14*(1), 195–200.

Prabhakar, S. (2001). *Fingerprint classification and matching using a filterbank.* Ph.D. thesis, Michigan State University, USA.

Shih, F. (2008). *Digital watermarking and steganography – fundamentals and techniques* (pp. 39–41). Boca Raton: CRC Press.

Xu, J., Pang, H., & Zhao, J. (2010). Digital image watermarking algorithm based on fast curvelet transform. *Journal Software Engineering & Applications, 3*, 939–943.

Yang, J., Hua, Y., & William, K. (2000). *An efficient LDA algorithm for face recognition.* Proceedings of the International Conference on Automation, Robotics and Computer Vision (ICARCV 2000), pp. 34–47.

Ying, L. (2005). *CurveLab2.1.2*. California Institute of Technology, USA.

Zhang, C., Cheng, L., Zhengding, Q., & Cheng, L. (2008). Multipurpose watermarking based on multiscale curvelet transform. *IEEE Transactions on Information Forensics and Security, 3*(4), 611–619.

Chapter 8
Conclusions and Future Work

Abstract This chapter presents a summary of the research work done which shows achievements toward mentioned objectives in the book. The major contributions of proposed research work are explained in this chapter. The comparison of proposed watermarking technique with existing watermarking technique is also explained in this chapter. Finally, some possible future direction is mentioned in this chapter.

8.1 Major Contribution of Proposed Research Work

The primary objective of this proposed research work is to provide security to biometric image against modification attack in a multibiometric system. The security of biometric image is provided by using compressive sensing (CS) theory and watermarking technique. This proposed watermarking approach added procedures such as CS theory acquisition procedure and reconstruction procedure in the conventional biometric watermarking approach. The acquisition procedure is added at the embedder side, while reconstruction procedure is added at the extraction side. Further, the aim of proposed research work is focused on providing security to watermark biometric image before embedding into host biometric image using CS theory. The proposed research work also aimed to study sparsity property of various image transforms such as Discrete Cosine Transform (DCT), Discrete Wavelet Transform (DWT), and Singular Value Decomposition (SVD). The major contribution of the proposed research is summarized in the following section. The proposed research work also evaluates against the proposed enhancement in the multibiometric system and multibiometric watermarking.

In this proposed research work, benefits of watermarking technique and CS theory are used for providing security to the biometric image. These proposed watermarking techniques provide security to biometric image against modification attack at system database and a communication channel between two modules in the multibiometric system. This book has also proposed the application of CS theory in the watermarking technique and security of biometric image.

© Springer International Publishing AG 2018
R. M. Thanki et al., *Multibiometric Watermarking with Compressive Sensing Theory*, Signals and Communication Technology,
https://doi.org/10.1007/978-3-319-73183-4_8

The various sparse watermarking techniques are designed and implemented in this book. These techniques used various image transforms such as Discrete Cosine Transform (DCT), Discrete Wavelet Transform (DWT), Singular Value Decomposition (SVD), and Fast Discrete Curvelet Transform (FDCT). The performance of proposed sparse watermarking techniques is evaluated using two quality measures such as PSNR and SSIM. By analyzing the results of proposed watermarking techniques, it is observed that the values of PSNR and SSIM are above 35 dB and 0.95, respectively, when any watermarking attack is not applied on watermarked face image.

The effect of proposed watermarking techniques on the performance of a multibiometric system is also analyzed in this book. The performance of the multibiometric system is evaluated using parameters such as the probability of verification, False Rejection Rate (FRR), False Acceptance Rate (FAR) and Equal Error Rate (EER). By analyzing the obtained results, it is observed that the performance of multibiometric system does not degrade due to these proposed techniques. The multibiometric system using proposed watermarking techniques can be used in high security applications such as ATM transactions verification, access control of laboratories, nuclear power stations, and the military base.

The comparison of proposed watermarking techniques is given in Table 8.1. The table shows that curvelet-based proposed watermarking techniques provided high-quality measures compared to other proposed watermarking techniques. The EER value of curvelet-based proposed watermarking technique is less than the EER value of other watermarking techniques for the multibiometric system. This indicates that the performance of curvelet-based proposed technique is better than all of other proposed watermarking techniques.

8.2 Comparison of Proposed Technique with Existing Technique

The comparison of proposed technique with existing technique is summarized in Table 8.2. This proposed technique is fragile against various watermarking attacks while existing techniques are robust. The security of watermarking biometric data before embedding is missing in existing technique. While in proposed technique, security of watermarking biometric data before embedding is provided using compressive sensing (CS) theory. The quality measure such as PSNR of proposed technique is higher than existing technique available in the literature.

Based on the comparison, the effect of watermarking on the performance of the multibiometric system is covered in this book which is not covered in many existing watermarking methodologies. The existing technique is used for protection of biometric data in the multibiometric system. The proposed technique is used for protection as well as authentication of biometric data in the multibiometric system.

Table 8.1 Comparison of proposed multibiometric watermarking techniques

Proposed multibiometric watermarking techniques	PSNR (dB) at $k = 0.2$	SSIM	Probability of verification at threshold value = 0.7	FAR at FRR fixed value = 0.7	EER
Using DWT	54.96	0.9865	0.909	0.069	0.301
Using DCT-DWT	43.62	0.9847	0.872	0.075	0.267
Using DWT-SVD	37.32	0.9504	0.966	0.052	0.264
Using FDCuT-DCT	64.07	0.9861	0.950	0.041	0.217

Table 8.2 Comparison of proposed technique with existing technique

Parameters	Existing technique	Proposed technique
Type of watermarking techniques	Many existing watermarking techniques are robust	One robust watermarking and three fragile watermarking techniques are proposed
Security of watermark biometric data before embedding	No such scope	Compressive sensing (CS) theory
Effect of watermarking technique for performance of multibiometric system	Missing in existing watermarking techniques available in the literature	Effect of proposed watermarking techniques to the performance of the multibiometric system is analyzed
Average PSNR value range	30–60 dB	35–65 dB
Application	Used for protection of biometric data over a nonsecure communication channel	Used for protection as well as authentication of biometric data over a nonsecure communication channel and a system database, respectively. Used in high security biometric-based applications

8.3 Future Research Direction

The application of CS theory in multibiometric watermarking is proposed in this book. The application of proposed techniques on face-fingerprint-based multibiometric system is analyzed in this book. In the future, these proposed watermarking techniques are applied to the security of other biometric data, such as iris, voice, signature, and palm print.

In the future, this CS theory can be applied to the area of image fusion, image registration, EEG signal processing, and ECG signal processing, because this data is easily converting into it's sparse domain when signal transform is applied on it. The real-time implementation of proposed techniques will be done using DSP kit in the future. The DSP kit such as TMS320C6000 with help of Code Composer Studio (CCS) platform will be used for real-time application of proposed techniques in the future. Also, these proposed techniques will be used for security of other multimedia data such as images, videos, and audio. These proposed techniques will be used in other applications such as security of data in mobile phones, secret document related to hardware, and so on.

Index

© Springer International Publishing AG 2018

R. M. Thanki et al., *Multibiometric Watermarking with Compressive Sensing Theory*, Signals and Communication Technology,
https://doi.org/10.1007/978-3-319-73183-4

Printed in the United States
By Bookmasters